FIREFLIES

AND OTHER

UAVs
(Unmanned Aerial Vehicles)

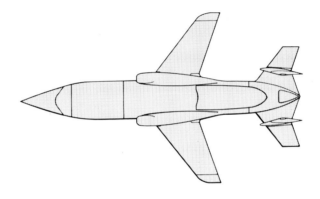

by WILLIAM WAGNER
and WILLIAM P. SLOAN

Published in 1992 by
Midland Publishing Limited
24 The Hollow, Earl Shilton,
Leicester, LE9 7NA, UK

by arrangement with
Aerofax, Inc.
Arlington, Texas 76006, USA

ISBN 1 85780 005 2

Manufactured in Hong Kong

To —

Hudson B. Drake

Robert A.K. Mitchell

G. Williams Rutherford

Robert R. Schwanhausser

without whose leadership and interest these UAVs — and this book — would not have been possible.

A WORD ABOUT BOOK TITLES

LIGHTNING BUGS or FIREFLIES?

Webster says the names for "night-flying beetles which emit phosphorescent light" are interchangeable — the same.

DURING WORLD WAR II, Ryan developed the FR-1 jet-plus-propeller Navy fighter aircraft which was given the popular name *Fireball*. At the same time, the Defense Department granted to Ryan future rights to the prefix *Fire* for subsequent military aircraft.

When Ryan won the tri-service Army-Navy-Air Force contract in the late 1940s for the first jet-powered aerial target, the name *Firebee* was selected. Remotely piloted vehicles bearing that name have been in constant production since then.

In the early '60s, intelligence-gathering reconnaissance versions of the *Firebee* were secretly developed under the name *'Fire Fly'* for possible use on military 'spy' missions over Cuba and later over Southeast Asia and Vietnam. Soon the code name *Fire Fly* was compromised and — logically — the classified identification became the alternative *Lightning Bugs*.

The name *Firefly* reappeared in 1969 when the Air Force found it prudent to disguise the existence and true code name of Ryan's Model 154 high-altitude, long endurance "Compass Arrow" unmanned vehicle, which had been developed for possible operation over China.

As a sequel to the author's definitive reconnaissance history, "LIGHTNING BUGS," it seemed appropriate that this updated account of what has since been accomplished should revert to the original code name for intelligence-gathering Unmanned Aerial Vehicles; thus . . .

"**FIREFLIES** AND OTHER **UAVs**"

William Wagner

San Diego, California
April 1992

CONTENTS

LIGHTNING BUGS

    Unmanned Reconnaissance . . . . 1—13

UAV BACKGROUND

    Joint Project Office . . . . 14—18

GOING SUPERSONIC

    Model 166, Firebee II . . . . 19—26

FLYING HIGHER and LONGER

    Model 154, Compass Arrow . . . . **27—47**

The INTERNATIONAL SCENE . . . . **48—74**

    ISRAEL'S 124I . . . . 48—64

    JAPAN — Targets for Defense . . . . 65—71

    NATO/SARDINIA . . . . 72—74

PILOTLESS TARGET AIRCRAFT

    U.S. Navy, Air Force, Army . . . . **75—93**

MULTI-MISSION UAVs . . . . **95—108**

    The BGM Series . . . . .98—108

LONG ENDURANCE/HIGH ALTITUDE

    Runway-based Compass Cope . . . .**109—118**

HIGH PERFORMANCE RPVs . . . .**119—131**

    FIREBRAND — Anti-Ship Missile Target . . . . .119—123

    FIREBOLT — Mach 4.3; 103,000 Feet . . . . .124—129

    SLAT — Supersonic Low Altitude Target . . . . .130—131

A FLOCK OF NEW BIRDS . . . .**132—141**

    MID-RANGE PREVIEW 133

    HIGH-FLYING 'SPIRIT' . . . . .132—134

    MINI-DRONES, Models 262, 326, 328 . . . . .135—138

    TARGETS FOR LASER WEAPONS . . . . .139—141

RETURN TO ACTIVE DUTY . . . .**142—150**

    OVER-THE-HORIZON . . . . .142—144

    NORTH WARNING RADAR . . . . .145—147

    'AEGIS' CRUISER TESTS . . . . .148—150

NEW GENERATION UAVs . . . . **151—193**

    THE AFFORDABLE 410 . . . . .151—156

    EGYPT'S 324 SCARAB . . . . .157—174

    BQM-145A UAV-MR . . . . .175—193

APPENDIX AND INDEX . . . .**194—204**

Most Teledyne Ryan unmanned aerial vehicles (UAVs) have military designations in the BQM, BGM and AQM series.

The <u>BQM</u> series (and <u>MQM</u>) are pilotless target aircraft used in weapons training and evaluation.

The <u>BGM</u> series are described as having 'surface attack' capability.

The <u>AQM</u> series are air-launched vehicles developed from basic Firebee designs. Most bear company Model 147 nomenclature. Note pages 12 and 13.

On the opposite page is a detailed tabulation of Teledyne Ryan UAV Models. Also see the indexed Model Directory on page 201.

Index Page 198 lists Firebee models and other unmanned vehicles having a "Fire" prefix.

ACRONYMS and ABBREVIATIONS Symbols	
A	Air Launch
B	Air and/or Ground Launch
G	Surface Attack
M	Missile
M (as first letter)	Mobile Launch
Q	Drone
X	Experimental
Y	Prototype
Z	Planning

TELEDYNE RYAN UAV MODELS
Tabulation covers key references to major production or research models mentioned in the text
(See other entries indexed under 'Models Directory')

Dates	TRA Model Number	User's Designation	Pages	User	Comments
1948—58	49 to 61	Q-2, KDA, XM-21 Series	15—16	U.S. Air Force, U.S. Navy, U.S. Army; also Canada	Firebee (Early models)
1959—	124	Q-2C, BQM-34A	75—92	U.S. Air Force, U.S. Navy	Firebee I
1963—	124E	MQM-34D	84—87	U.S. Army	Firebee I
1964—75	124 SERIES	AQM SERIES	1—13	U.S. Air Force	Fire Fly (1962) Lightning Bugs
1970—	124I	124I	48—64	Israel	Yom Kippur War
1968—	124	BQM-34AJ	65—71	Japan	Ship-Launched from 'Azuma'
1984—	124RE	(RE = Arab Republic of Egypt) See Model 324			
1984—	124	BQM-34S	87, 91—93	U.S. Navy	Firebee I with ITCS
1970—	147SC/SD	AQM-34L/M	142—150	U.S. Air Force	OTH; NWR; Aegis
1966—71	154	AQM-91A	27—47	U.S. Air Force	Compass Arrow (Firefly)
1966—90	166	BQM-34E/F	19—26	U.S. Navy, U.S. Air Force	Firebee II (Supersonic)
1973—	232	232	64, 72—74	Israel, NATO/Sardinia	International 124 Targets
1971—72	234	BGM-34A/B	98—101	U.S. Air Force	Multi-Mission/Strike
1972—76	235	YQM-98A	109—117	U.S. Air Force	Compass Cope-R
1971	248	BQM/SSM	96—98	U.S. Navy	Weapons Delivery
1973	251	MQM-34D MOD II	86—87	U.S. Army	With J-85 Engine
1973	255	AQM-34V	102—105	U.S. Air Force	Combat Angel (ECM)
1977—79	258	ZBQM-111A	119—123	U.S.Navy	Firebrand
1974—78	259	BGM-34C	105—108	U.S. Air Force	Multi-Mission
1975—85	262 (and 326, 328)	Miscellaneous	135—138	Various	Mini-Drones
1979—84	305	AQM-81A/N	124—129	U.S. Air Force, U.S. Navy	Firebolt (HAHST)
1987	314	Sealite	139—140	U.S. Navy	Target for Laser Research
1984—	324	(124RE)	157—174	Egypt	'Scarab'
1986—90	336	'HATS'	140—141	U.S. Navy	Laser Beam Research
1988—	350 POC	MR-UAV	176—181	Teledyne Ryan	Proof-of-Concept
1989—	350	BQM-145A	176—193	U.S. Navy, Marines, Air Force	
1985—	410		151—156	Teledyne Ryan	'Affordable' RPV
1988—	410'B'	(See 350 POC)			

INTRODUCTION

UAVs — UNMANNED AERIAL VEHICLES. That's the newest terminology for the growing family of pilotless aircraft and winged missiles formerly referred to as 'drones.'

Drones made their first significant appearance in the early fifties soon after Ryan Aeronautical Company was declared winner of a tri-service Army-Navy-Air Force competition for a jet-propelled aerial target system.

For fifteen years the emphasis remained on the capability of drones as targets. Then, during the war in Vietnam, some were modified for 'spy' missions over enemy territory. That opened up a whole new field for drones.

Today UAVs fall into three main categories:

1) **Pilotless target aircra**ft (PTAs) — used to train personnel in air-to-air and surface-to-air target practice, and for testing the effectiveness of new weapons.

2) **Reconnaissance vehicles** which gather intelligence information, usually over 'enemy' territory. The role of such remotely piloted vehicles (RPVs) is not threatening in nature as their purpose is to improve the capability of a sovereign country to defend itself against an enemy.

3) As **Weapon delivery systems** to take the offensive against an aggressor with lethal military strikes. (Cruise missiles, having airfoils, are also a good example)

The key to the mission role is the 'payload' carried by the UAV. As a PTA, the payload could include flares to simulate the tail pipe of an enemy jet fighter; or in an RPV, cameras to photograph enemy installations; or as a delivery system, carrying a lethal bomb to be air launched in a strike against enemy installations.

Mission requirements mandate a wide range of operational capabilities. For example:

- Must the unmanned vehicle be able to fly for long periods at high altitudes, serving as a **reconnaissance** platform?

- Does the **target** mission require flying at supersonic speeds at low altitude, simulating an enemy missile attacking a naval ship?

- Is the bird capable of carrying heavy loads suspended beneath its wing and **delivering weapons** in an attack on an enemy installation?

Some flights can be multi-missions, combining gathering real-time visual information with a TV camera, then launching a missile from the same UAV to attack the enemy installation.

In each category, there are many different applications of technology. For example, a non-lethal RPV can carry out a wide variety of intelligence missions such as photography, air sampling, electronic eavesdropping, flying as decoys to confuse the enemy, and target spotting, to mention some of many uses.

ONE OF THE PRIME ADVANTAGES demonstrated in Vietnam was the ability to fly unmanned reconnaissance routes over enemy territory without exposing human pilots to such dangerous missions.

As Air Force Maj. Joe Tillman later explained, "The RPVs, with nerves of steel, can get in, do their thing, and get out. If an RPV doesn't make it, we've lost a 'vehicle,' not a flight crew and a multi-million dollar aircraft."

For a time, RPVs suffered a poor image among older pilots who saw them as a challenge to their individuality. As one pilot asked, "How can you be a tiger, sitting behind a ground-based console?"

To that, Tillman's response was, "Any pilot who wants to fly the typical RPV profile in a high-threat combat area needs to have his head examined."

TO UNDERSTAND SOME OF UAVs' more exotic uses one need only refer to the definitive text, 'LIGHTNING BUGS,' which described their role in Vietnam on spy missions.

During and since those pioneering unmanned reconnaissance flights, Teledyne Ryan Aeronautical has been on the cutting edge of new technology in the UAV field. Thus, this book devotes itself not only to work done in the two decades since Vietnam, but also to some of the projects under way even while primary attention of necessity was being given to events in Southeast Asia.

But, first, the following review of RPVs' 1964-1975 operational record in Asia will be useful. It is condensed from LIGHTNING BUGS (published 1982 by ARMED FORCES JOURNAL International).

68 SORTIES

Above   **'LIGHTNING BUG' INSIGNIA** *worn as a shoulder patch by Ryan technical personnel serving in Vietnam.*

Preceding page **RECORD-HOLDING RECONNAISSANCE DRONE** *was 'Tom Cat,' a 147SC model, operated by the 99th Strategic Reconnaissance Squadron out of U-Tapao Royal Thai Airfield. 'Tom Cat' flew 68 sorties with an average of 12 targets covered per mission.*

Photo by Dave Gossett, Teledyne Ryan Aeronautical.

The Saga of

UNMANNED RECONNAISSANCE

WHEN FRANCIS GARY POWERS parachuted from his U-2 reconnaissance plane over Russia May 1, 1960, and was held captive for many months, the need for operating unmanned 'spy' missions was evident.

Only seven months earlier, in September 1959, the Chief of the Reconnaissance Division, U.S. Air Force, observed that "someone had better be giving some thought to the problem we're going to have if and when a U-2 pilot comes down in unfriendly territory."

Three years later, on October 30, 1962, Major Rudolph Anderson Jr., flying a U-2 over Cuba, where Russian missile installations had been sighted, was also shot down. But, in the intervening three years, programs for unmanned jet flights were already under way.

The metamorphosis of the Ryan 'Firebee,' a jet-powered recoverable aerial training target, into the vital role it played as a reconnaissance drone in the Vietnam conflict began long before that struggle involved the United States.

It was in early 1960 that Robert R. (Swany) Schwanhausser, then project engineer for Ryan Aeronautical Company, made his first pitch to the Air Force Reconnaissance people at the Pentagon. Following a brief history of the capabilities of the basic Firebee training target he gave a classified talk, outlining the possibility of utilizing unmanned aerial drones with a range capability of 1200 to 1400 nautical miles for intelligence gathering. While his viewgraphs were based on theoretical flights within the U.S. the potential coverage of foreign areas was not overlooked.

Despite the grim fact that the Air Force had little money, no line items or budget for such a program, Ryan managed to obtain a token order for $200,000 for a feasibility and integration study. With the budgetary purse strings loosened a tad, Ryan's 'Red Wagon'

program got underway in 1960, followed by 'Lucy Lee' in 1961.

A forerunner of 'stealth technology' to reduce the radar reflectivity of aircraft made its debut at this time in early Red Wagon flight tests. A screen over the engine air intake, nonconductive paint and radar-absorbing blankets on sides of the airframe reduced detectability of the modified **Q-2C** test Firebee.

The projected Ryan **Model 136** would further reduce detectability by placing the jet engine atop the fuselage to reduce the drone's infrared signature.

Both the Red Wagon and Lucy Lee programs were eventually terminated because of the developmental costs involved. But, after two years of frustration, the 'Big Safari' quick reaction management concept appeared on the horizon, and its utilization by the Air Force and Ryan culminated in a successful future for unmanned 'recce' aircraft.

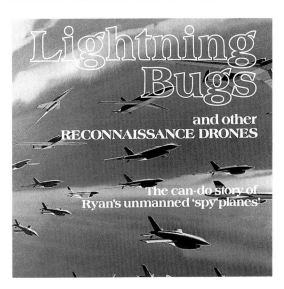

Lightning Bugs and other RECONNAISSANCE DRONES

The can-do story of Ryan's unmanned 'spy' planes

BIG SAFARI WAS A SYSTEM of procurement for special reconnaissance that had survived since the early '50s. It was an expedited means of bypassing the old Research and Development system, providing a rapid response capability. This made it possible for military and contractor to expedite every phase of a program, eliminating endless red tape. While its original intent was to expedite the modification of manned aircraft for recce missions, there were no directives prohibiting its use for unmanned aircraft.

Col. Lloyd M. Ryan of the Reconnaissance Division uncovered this gem from the archives, and with Swany's help secured a contract for modifying four standard Q-2C Firebee training targets into recce birds. This 'Fire Fly' program resulted in a test and evaluation of the newly designated Ryan **147A** birds in New Mexico, followed by five proving flights off the coasts of Florida (with all of the attendant trials and tribulations of off-site operation).

With three surviving 147 Remotely Piloted Vehicles (RPVs) remaining, the Strategic Air Command (SAC) recommended that those birds be retired and not used in the Cuban missile crisis. However, SAC came up with enough money to update more Q-2Cs for future use.

Fortunately, Ryan and the Air Force were well prepared for the Vietnam crisis, in large part because of the role played at the Joint Chiefs of Staff level by then Col. R.D. (Doug) Steakley, its reconnaissance director.

The code name Fire Fly having been compromised, the recce program then took on the 'Lightning Bug' classified identification.

LEADERS OF THE 'BIG SAFARI' were Air Force *Col. Lloyd Ryan, left, and Ryan's Bob Schwanhausser. They sparked the 'can-do' quick reaction management style that made 'Lightning Bugs' glow.*

BY THEN THE BASIC Q-2C design had been configured for recce missions. Wings, extended from 13 feet to as much as 27 feet greatly increased altitude capability. Extending the fuselage and nose sections provided space for more fuel to increase operational range, and greater payload for sophisticated intelligence-gathering equipment. And, while Q-2C training targets could be ground or air launched, the 147s were configured only for launch from converted C-130 cargo planes.

IT WAS AUGUST 10, 1964 when President Lyndon B. Johnson issued his mandate — the Tonkin Gulf Resolution — and the United States became involved in a 'police action' (no-win war) in which unmanned aircraft were to play a vital part.

The Strategic Air Command, realizing that a group of trained Air Force and Ryan 'can-do' people were available, set in motion the deployment of experienced personnel and **147 Model B** birds to the Kadena Air Base in Okinawa. Flights were made along the coast and interior of southeastern China with parachute recovery on Taiwan. Having proved the potential of RPVs to be a useful instrument in aerial reconnaissance, the group then redeployed to Bien Hoa Air Base, 20 miles north of Saigon in South Vietnam, for expanded operational use.

Recognition must be given to the contractor's 'First Team' in Vietnam - Ryan technicians with at least ten years experience in flying, maintaining and operating RPVs. The group included such stalwarts as Ed Sly, Billy Sved, Jack Lucast, Gene Motter and Dale Weaver, and tough 'Big Jim' Regis sent by the Air Force as a civilian technical specialist. They were later followed by Bob Schwanhausser and Air Force types who backed up the overseas operation for the duration.

MISSION VERSATILITY BY AIRFRAME MODIFICATION

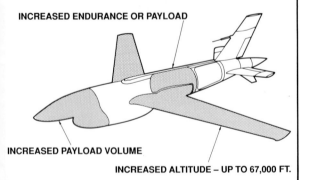

INCREASED ENDURANCE OR PAYLOAD

INCREASED PAYLOAD VOLUME

INCREASED ALTITUDE – UP TO 67,000 FT.

FOUR DECADES EARLIER Ryan extended the wing *of its new monoplane design to get Charles A. Lindbergh nonstop to Paris in the "Spirit of St. Louis." – Ed.*

A 147D IS LAUNCHED from DC-130. 'Stretched' version of Firebee target was four feet longer and wing span two feet wider. In 147B, wing span was widened 12 feet more.

In mid-November 1964, the Chinese Communists announced that a "pilotless high-altitude reconnaissance military plane of U.S. imperialism" had been shot down over South China. Soon after, photos of the downed Ryan 147**B**, on display in Peking, appeared in world-wide newspapers. In the tradition of secret intelligence operations, no confirmation came from U.S. sources, but the 'spy plane' missions continued.

THE STORY OF UNMANNED vehicles in Vietnam is little understood by those not closely associated with their participation in providing the Defense Department with vital information, including electronic intelligence gathering, surveillance, photo reconnaissance and other tasks that saved the lives of scores of pilots of manned aircraft.

A total of 3435 operational reconnaissance sorties were flown in Southeast Asia between 1964 and 1975, and involved a total of 1016 of the 147 vehicles of varying configuration and model designations. Each model, each mission flown would provide an endless story of ingenuity, pathos, heroism and humor.

The first birds deployed were the big-wing high-altitude day photo 147**B** series. Capable of flying at altitudes in excess of 50,000 feet, their ability to return with pictures of high-resolution photography covering areas denied by political edict to manned aircraft was invaluable. Seventy-eight missions were flown with these birds.

Flights over North Vietnam and southeast China were air-launched from the four-engine DC-130 cargo planes, with parachute recovery north of Saigon, at Da Nang.

FIRST DRONE SHOT DOWN by Chinese was a 147B lost over south central China on November 15, 1964.

Air-launching gave the spy missions 'longer legs,' adding the range of the mother plane to that of the recce bird.

Several new Ryan reconnaissance models were soon being developed, including the short-wing, low-altitude **147C** and **D** for gathering electronic intelligence. The longer-wing **E** and **F** models were for high-altitude electronic intelligence and for electronic countermeasures.

A major step-up in performance came with the **147G**, the first model to be powered by an updated J-69 Continental CAE jet engine producing 1920 lbs. thrust in place of the 1700 lbs. version which powered earlier models. The G's wing area was 80 sq. ft., same as the **B**, but double that of the earliest Ryan recce birds.

For still higher altitude, longer range photo missions, the **147H** model began operation in March 1967. It boasted a 32 foot wing span and an area of 114 square feet.

Ten standard **BQM-34A** training targets were switched over to a new series of **147N** models to be used as expendable decoys preceding **G** bird missions and, later, in similar roles for the **J** birds.

Recce flights 'up north' were conducted under command of Col. Bill Forehand for the first six years from Bien Hoa in South Vietnam, then on July 4, 1970 the 100th Strategic Reconnaissance Wing, under command of Col. John Dale, moved operations to the Royal Thai Airfield at U-Tapao, Thailand.

METEOROLOGY PLAYS an important role in the conduct of any conflict. Overcast skies prevent the high-flying recce planes from peeking underneath and a low cloud cover requires the capability of seeing what's underneath. When SAC asked for help, the big-wing **147J** model was configured for low-altitude work.

The high-altitude birds had to face the twin hazards of frequent attacks by SA-2 missiles and by the Russian-built MiG fighters. A hurried R&D test program was instituted at the Navy's facility at Pt. Mugu in Southern California. While the J program suffered a few minor catastrophies the field service team perfected BLACS, a Barometric Low Altitude Control System, which enabled the vehicle to automatically descend from 20,000 feet to under 1,000 feet.

Within two months, the **J** bird was revealing installations previously unseen with the 'undercast' conditions. However, use of this stretched-wing configuration was similar to training an eagle to fly like a sparrow-hawk, and its usage was limited. It did provide basic engineering knowledge for its successors, the **147S** series.

"Firebee training targets," relates Dale Weaver, "had always been recovered by parachute at the end of their missions and retrieved for reuse. But in Vietnam the **147**s were getting badly clobbered as they parachute-landed into rice paddies, jungles and the ocean off Da Nang, our recovery base."

Back in San Diego, Ryan technicians led by former Col. Fred Yochim came up with a better way — the Mid-Air Retrieval System (MARS). After parachute recovery was initiated, a helicopter could snatch the descending bird in mid-air and return it undamaged to the base at Da Nang. MARS' first use was with the 147**J** bird.

During a later period, the Navy was trying out aircraft carrier-launched **147SK** birds and didn't like to recover them from the salty ocean because it was necessary to decontaminate them before reuse. If the Air Force had a chopper available at Da Nang it helped out with a MARS recovery. On one occasion after delivering a dry **SK** to the deck of the Navy carrier 'Ranger,' the Air Force helicopter pilot stepped out carrying an American flag declaring "I claim this 'island' for the United States Air Force."

A VALUABLE ASPECT of the conflict was the relationship that existed between the military and civilian personnel.

Little or no animosity existed. Instead there was friendly competition as to who could

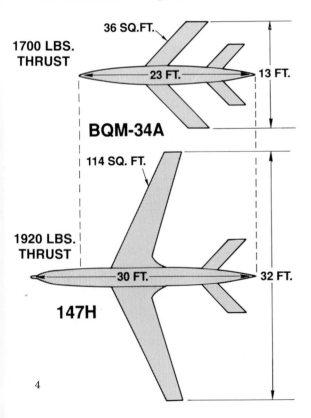

36 SQ.FT.

1700 LBS. THRUST

23 FT.   13 FT.

BQM-34A

114 SQ. FT.

1920 LBS. THRUST

30 FT.   32 FT.

147H

Dave Gossett

SNATCHED IN MID AIR by a helicopter, *a 147H recce bird is flown back to base for another launch from its DC-130 mother plane.*

do the job better, a complete exchange of technical information, and an infectious can-do attitude that transcended the red tape restrictions of an exclusively military operation.

Ryan personnel were as much in a combat situation as their GI counterparts. Mortar shells, bombings and hazardous duties were part of their mission, which they shared with the military ground and air support people. Heroes there were, but no decorations, no purple hearts for injury; instead a tremendous amount of satisfaction for getting the job done.

One example was the performance of Ed Christian. Ed, a Ryan camera systems and photo interpretation expert, was lowered from a helicopter in an enemy-held jungle to retrieve the film from a downed drone. Rescued from enemy sniper fire by a circling support chopper, he returned to base with the film intact.

Direct Control Operators (DCOs), along with the Air Force personnel aboard the C-130s, which launched the drones, were subjected to SAM missiles and MiG intercepts. Fortunately, there was not a single fatality recorded by contractor personnel during the years they spent in Vietnam.

NO MISSION FLOWN could be classified as 'typical.' The prime objective of each flight was to successfully launch the bird from a C-130 aircraft at the proper time and the proper place and put it through the available 'window.' Aboard the launch aircraft were the Air Force airplane drivers and the direct control military and civilian personnel who checked out the internal electronic programming functions and got the drone off the wing and on its way.

If the bird did its job properly, it would return to a predesignated location, pop a pair of parachutes and be picked up in mid-air by an Air Force helicopter to be returned for sensor evaluation.

Dale Weaver described the routine in regular messages back to Ryan headquarters:

"We were to launch in the Gulf of Tonkin 15 miles off Haiphong which took us within 22 miles of a new SAM site. We went charging up the Gulf and could hear our fighter CAP (combat air patrol) overhead but they couldn't find either us or their own aerial refueling tanker.

"As we started in the bird had a minor autopilot malfunction. Then to make things more interesting we received an indication from the U.S. 'spook' planes that our C-130 launch plane was being painted by radar from an SA-2 missile site. As we turned in to launch we looked pretty hostile to them and I imagine if we'd gone much closer they would have fired their SA-2 on us.

"We had already gone through an aborted launch because of the autopilot malfunction and the crew knew a SAM had locked on us

"YOU SHOULD SEE ME NOW! I am truly *the front line combat troop, go-to-hell hat and all." That's Dale Weaver at left with Capt. Charles West, DC-130 navigator.*

Bruce Engman

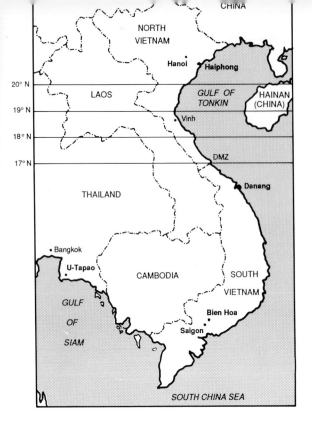

THOSE INVOLVED IN THE RECCE program — whether Air Force or contractor personnel — were, as a group, deeply involved and committed to the success of unmanned reconnaissance.

Col. Lloyd Ryan had certainly done his job well. From late 1959 to the end of 1963 he, more than anyone else within the service, had kept drone reconnaissance alive. He had done the missionary work.

Another outstanding crusader for the cause was Col. Ellsworth Powell. He was a triple-threat player in the game of drone survival. Ells understood the mechanical side of the drone picture — he was an expert in the operations end of the business and so fully understood the procurement side at the Air Force Logistics Command.

Team these leaders, including the likes of Colonels Andy Corra, John Dale and Bill Forehand, with their industry counterparts and you had an outstanding team.

The chief driving force at Ryan was Bob Schwanhausser, supported by such engineering and administrative assistants as Erich Oemcke and Bob Reichardt.

After studying aeronautical engineering at M.I.T., Schwanhausser's first contact with target drones was as an Air Force Lieutenant assigned as project engineer on Ryan Q-2 Firebees in the early days at Holloman Air Force Base in New Mexico.

When Lt. Schwanhausser completed his tour of duty he was plucked off by Continental Aviation and Engineering (CAE) as project engineer on jet engines which powered Ryan drones. Shortly thereafter Swany was employed by Ryan. Twenty-nine years later he left Teledyne Ryan Aeronautical to become President of Teledyne CAE.

and was in the fire mode. It wasn't exactly the safest area in which to fly a holding pattern. Things were getting a little sticky and at a very critical stage."

After the subsequent launch, the crew reported that Dale had remained pretty cool except for a "damn it, make another orbit, and I'll get the son-of-a-bitch off." Which he did. And right over the North Vietnamese fishing fleet.

Typical mission? Well, not your ordinary one, but indicative of some of the problems in getting a bird on its way.

The average life of all of the RPVs launched in Vietnam was around three and a half missions per bird. However, there were many that survived an incredible number of sorties. Champion of the flock was 'Tom Cat' a **147SC** recce bird which set the outstanding record of 68 missions before being lost.

An idiosyncrasy of Americans, to have an affinity with the machines they work with, is borne out in the names of those drones that survived. 'Budweiser' (probably named by a master sergeant) racked up a score of 63. 'Ryan's Daughter' came home with 52 successful sorties into enemy territory, and 'Baby Buck' (Col. Buck Lee's favorite) scored 46 successful missions. And the 99th Strategic Reconnaissance Squadron flew over 100 consecutive missions without losing a single bird.

AIR FORCE COL. JOHN DALE and his stellar SC *performer 'Tom Cat,' the all-time record setter with 68 successful missions to its credit.*

Ryan Aeronautical Library

Photo at right: **CLOSE UP OF POWER TOWER** *was taken* by *147SB-12 on 6 October 1968 as it flew beneath high tension lines.* Below: **RED CHINESE MILITIA celebrate** *shooting down of 147H-18.*

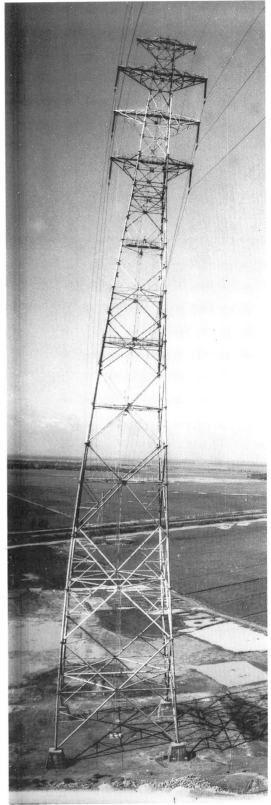

Ryan Aeronautical Library

Protected by a Top Secret security blanket, both military and civilian officials were able, when needed, to go directly to top government decision makers, often bypassing generals and cabinet members to obtain rapid action. But none ever lost sight of the prime objectives of furthering the war effort by conducting unmanned 'spy' flights which not only gathered strategic information but also saved the lives of American pilots.

Where 'quick reaction' management was the name of the game, the atmosphere was often like a pressure cooker. It was hard to keep the lid on!

The writer (Sloan) was Ryan's program manager for the 147SC series of low-level birds, the largest and least costly production run of any recce model. In 'Lightning Bugs,' author Bill Wagner quotes me in this way:

"You couldn't help but feel that the work you were doing was vital to the national interest. We figured if our birds were going in and coming back with the pictures that were saving lives of American pilots and others who might be in jeopardy, then doing our jobs right was truly a matter of life and death.

"So when those of us in the program office ran into anyone in the plant or military who was grumbling about having to work overtime, or dragging his feet, or saying it can wait until tomorrow, your blood pressure got higher and higher." [Sloan had his heart attack soon after that, followed by Schwanhausser six months later! — Ed.]

The 147**S** series flew a total of 2,369 missions, of which 1,651 were made with the 147**SC** (AQM-34L), the low altitude workhorse. It was Big Safari's Col. Walt Raynor who worked with Ryan's Bob Reichardt to get the **S** program up and going.

U.S. Air Force

7

WITH G AND H BIRDS on the launch racks, *the DC-130 heads out from Bien Hoa on missions to be flown 'up north' by the unmanned reconnaissance drones.*

DETAILS OF ENEMY OPERATIONAL AAA battery *were also photographed 6 October 1968 by low-altitude 147S series RPV.*

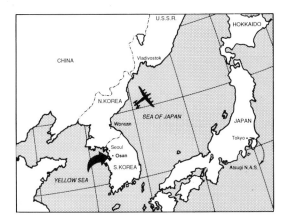

IN JUNE 1969, an Air Force Super-Connie, designated as EC-121, was shot down off the North Korean coast with 31 men aboard. Its mission was communication and electronic intelligence gathering, and again this incident created the need for an unmanned vehicle to do the same job.

Briefly, the mission of these flights was to listen to verbal radio transmissions by the enemy, detect the frequencies of the SAM missile launch sites and guidance systems of the missiles, transmit that information to B-47 aircraft well out of target range and disseminate that information to the defense centers throughout the theater.

Six months after the EC-121 was shot down, the new higher performance Ryan **147T** model became operational. It boasted a newly developed Continental CAE jet engine of 2800 lbs. thrust, a 45% increase over the engine in the **H** high-altitude bird.

A third operational base, for its use, was established at Osan, South Korea for flights which permitted the on-board electronic-snooping gear to listen in on North Korean, Chinese and Russian communications and radar.

These 'ear in the sky' flights, offset at sea to cover the Demilitarized Zone (DMZ) which separated the two Koreas, continued for two years after the Vietnam cease fire of January 27, 1973.

The high-altitude **TE** and **TF** models were capable of intercepting signals from target transmitters at ranges up to 600 miles, with a relay system to transmit received signals back to U.S. ground stations in real time.

The basic **147T** photo reconnaissance models operating in the Vietnam theater and off the coast of Red China were responsible for saving scores of pilots, both in obsoleting the manned spy planes and in informing the fighter jockeys when the SAMs were being fired at them.

Shortly after the **Ts** were operational, the bad guys began trying to shoot them down. Borrowing evasive tactics from manned aircraft, operation 'Chicken' was introduced, allowing the drone to maneuver right or left when being approached by a SAM. On several occasions, the SA-2s nearly succeeded, with near-misses. One recce bird was even able to photograph the enemy missile bursting only a few yards away.

In all, with the ability to maneuver out of harm's way, the birds outguessed eight MiG intercepts, three air-to-air missile launches and nine ground-to-air launches. Smart drones were finally out-smarting the smart missiles!

147TF CARRYING TWO EXTERNAL FUEL TANKS for extended range missions is launched on flight out of Osan to monitor electronic activity along the DMZ.

LESS EXOTIC MISSIONS were also part of the operation in Southeast Asia. Drones were called on in July 1972 to play a role in a stepped-up propaganda war against North Vietnam. The probability of a manned aircraft flying deep into enemy territory to drop leaflets and return undamaged was almost nil.

Several Model **147NC** drones were modified with external pods, generally used for chaff dispensing, but now containing leaflets with a message from President Nixon asking the North Vietnamese to please quit this foolishness. The success of these missions was not earth-shaking, and while known as project 'Litter Bug' in more polite circles, these NC drones were referred to by the operational troops as 'Bullshit Bombers.'

BROAD WING 147H had 2,400 mile range, *altitude capability of 70,000 feet and new camera able to scan area 780 miles long and 22 miles wide.*

Dave Gossett

The prime usage of the 147 RPVs in Vietnam was surveillance. Photographs taken from either high or low altitude, with details so clearly decipherable that minute objects could be detected, played a significant role in that conflict. Precise location of missile emplacements, ship activity in Haiphong Harbor (including cargo being carried in them), enemy airfield locations, and bomb damage assessment photos rendered intelligence otherwise unobtainable, unless manned aircraft, with extreme risk to pilots, were used.

RESULTS OF PRE-CHRISTMAS 1972 'Linebacker' B-52 bombing *raids on Hanoi Power Plant. Photos shown Military Target Area 2 before and after bombing.*

HANOI THERMAL POWER PLANT

BEFORE

Department of Defense

HANOI THERMAL POWER PLANT

AFTER

Department of D

ONE OF THE GREAT psychological lifts for some of the American POWs in Hanoi was the sight of our low-flying birds swooping over the 'Hanoi Hilton.' Returned POWs were joyous in describing the flights. One reported, "Sometimes we heard the drone. Sometimes we saw it. After awhile the usual comment was 'there goes the little guy'." Another remarked, "I was standing out in the open in the middle of the compound when a drone approached overhead. I figured it was taking pictures, so I just stood up and smiled for the camera, hopeful that somebody back home might recognize me!"

A five-and-a-half year guest of the Hanoi Hilton was Commander Edward H. Martin, executive officer of VA-34 attack squadron from the carrier 'Intrepid.' He was approaching Hanoi in an A-4C, and after successfully eluding several SAMs, flew into a burst that put him out of action. He was interned, and although subjected to extreme torture, managed to survive the conflict.

His history, told later at the Ryan plant in San Diego, was a tribute to the drones' participation from a POW standpoint. Having taken part in Navy shoot-outs against Firebee targets, he was able to recognize the low-flying birds. Part of this history:

"We saw many recce drones during their intelligence-gathering flights over the Hanoi-Haiphong area. They were the only aircraft we heard for a long, long time and about the only thing that did lift our morale during those years.

"When the Air Force started their awesome, devastating bomber strikes in mid-December 1972 we knew we were coming home. By then the low-level 147 drones were flying their daylight bomb assessment sorties.

"The last flight we saw was about January 20, 1973 and it was indeed impressive as it split the prison compound right down the center and flew off to the southwest. We were waving and shouting to it because we always did what we could to let our presence there be known."

[Seventeen years later Vice Admiral Ed Martin retired as Deputy Chief of Naval Operations (Air). His familiarity with unmanned aircraft will no doubt have had a positive influence on the future of UAVs. — Ed.]

COMDR. ED MARTIN ON A VISIT to Teledyne Ryan *after return of American POWs meets with Adm. Ulysses S. Grant Sharp (USN, Ret.) and Robert C. Jackson, Teledyne Ryan chief executive. Adm. Sharp was Commander in Chief Pacific (CINCPAC) 1964-68.*

THE RECORD OF UNMANNED vehicles in Vietnam is a saga of men and machines that opened the way years later for the global employment of these systems for other defense purposes. At the termination of hostilities, the Ryan troops were returned to the U.S. — not to retirement, but to continue to play an active role in the future of UAVs in maintaining leadership in that field.

What was to become Ryan Aeronautical Company was founded in 1922 by T. Claude Ryan. After 46 years as an independent, publicly held corporation, Ryan was purchased by Teledyne, Inc., in 1969, while 'Lightning Bug' reconnaissance RPVs were operating in Southeast Asia. The successor company is now known as Teledyne Ryan Aeronautical (TRA).

TELEDYNE RYAN AERONAUTICAL
FAMILY OF UNMANNED AERIAL VEHICLES

AQM-34N
147 NX
147 NRE
AQM-34K
COMPASS BIN
YQM-98A
COMPASS COPE
BGM-34A/B
147 G/J
AQM-34M
(EXTENDED RANGE)
COMPASS BIN
BUFFALO HUNTER
BQM-145A
UAV-MR
147 E
147 NQ
AQM-34M
COMPASS BIN
BUFFALO HUNTER
AQM-91A
COMPASS ARROW
147 D
AQM-34L
COMPASS BIN
BUFFALO HUNTER
AQM-34R
COMPASS BIN
COMBAT DAWN
MODEL 410
147 B
147 NP
147 SA/SB
AQM-34Q
COMPASS BIN
MODEL 324
147 A/C
AQM-34H
COMPASS BIN
COMBAT ANGEL
BQM-34A/MQM-34D
FIREBEE
AQM-34G
COMPASS BIN
COMBAT ANGEL
AQM-34J
COMPASS BIN
COMBAT ANGEL
BQM-34E/F
FIREBEE II
AQM-34P
COMPASS BIN
AQM-81
FIREBOLT

RYAN RECONNAISSANCE MODEL DIRECTORY

Ryan 147 Model	Military Model	Length '	Span '	Area □ '	Thrust (lbs.)	Mission	Month/Year Operated	Number Launched	% Return	Most Flights by a Bird
A		27	13	36	1700	Fire Fly — first recce demo drone	4/62-8/62			
B		27	27	80	1700	Lightning Bug — first big-wing high-altitude day photo bird	8/64-12/65	78	61.5%	8
C		27	15	40	1700	Training, and low-altitude tests	10/65			
D		27	15	40	1700	From C for electronic intelligence	8/65	2		
E		27	27	80	1700	From B for high-altitude electronic intelligence	10/65-2/66	4		
F		27	27	80	1700	From B — electronic countermeasures	7/66			
G		29	27	80	1920	Longer B with larger engine	10/65-8/67	83	54.2%	11
H	AQM-34N	30	32	114	1920	High-altitude photo; more range	3/67-7/71	138	63.8%	13
J		29	27	80	1920	First low-altitude day photo (BLACS)	3/66-11/67	94	64.9%	9
N		23	13	36	1700	Expendable decoy (from BQM-34A)	3/66-6/66	9	0	
NX		23	13	36	1700	Decoy and medium-alt. day photo	11/66-6/67	13	46.2%	6
NP		28	15	40	1700	Interim low-altitude day photo	6/67-9/67	19	63.2%	5
NRE		28	13	40	1700	First night photo (from NP)	5/67-9/67	7	42.9%	4
NQ		23	13	36	1700	Low-altitude NX; hand controlled	5/68-12/68	66	86.4%	20
*NA/NC	AQM-34G	26	15	40	1700	By TAC for chaff and ECM	8/68-9/71			
NC	AQM-34H	26	15	40	1700	Leaflet dropping (Bullshit Bombers)	7/72-12/72	29	89.7%	8
NC(M1)	AQM-34J	26	15	40	1700	Interim low-altitude, day photo and for training				
S/SA		29	13	36	1920	Low-altitude day photo	12/67-5/68	90	63.3%	11
SB		29	13	36	1920	Improved SA low-altitude bird	3/68-1/69	159	76.1%	14
SRE	AQM-34K	29	13	36	1920	Night photo version of SB	11/68-10/69	44	72.7%	9
SC	AQM-34L	29	13	36	1920	The low-altitude workhorse	1/69-6/73	1651	87.2%	68**
SC/TV	-34L/TV	29	13	36	1920	SC model with real-time TV	6/72-	121	93.4%	42
SD	AQM-34M	29	13	36	1920	Low-altitude photo; real-time data	6/74-4/75	183	97.3%	39
SDL	-34M(L)	29	13	36	1920	SD bird with Loran navigation	8/72	121	90.9%	36
SK		29	15	40		Navy operation from aircraft carrier	11/69-6/70			
T	AQM-34P	30	32	114	2800	Larger engine; high-alt. day photo	4/69-9/70	28	78.6%	
TE	AQM-34Q	30	32	114	2800	High-altitude; real time Comint	2/70-6/73	268	91.4%	34
TF	AQM-34R	30	32	114	2800	Improved long-range TE	2/73-6/75	216	96.8%	37

3435 operational sorties by 100th SRW

NOTE: * NA/NC Combat Angel birds were operated on standby in U.S. by Tactical Air Command for possible pre-strike ECM chaff-dispensing missions.

** 68 missions by Tom Cat
63 missions by Budweiser
52 missions by Ryan's Daughter
46 missions by Baby Buck

BACKGROUND

PAST
The 1950s

Don Doerr

ABOVE - INITIAL GROUND LAUNCH OF Q-2 FIREBEE four decades ago *took place at Holloman Air Force Base in New Mexico using 11,000-pound JATO booster and 90-foot rail launcher.*

BELOW - INITIAL AIR LAUNCH OF MODEL 350 Proof-of-Concept *Mid-Range UAV was made from an F-4 jet fighter operating above the California desert from Mojave, adjacent to Edwards Air Force Base.*

and PRESENT
The 1990s

John Ligon

In 1948, Ryan Aeronautical Company, a pioneering aircraft company founded a quarter-century earlier, was declared winner of a tri-service Army-Navy-Air Force competition for a jet-propelled pilotless aerial target system.

But the story of unmanned air vehicles can be traced back much farther – to the fourth century B.C.

IT WAS OVER two thousand years ago that a young man, stood on a lonely wind-swept hill in China and flew recorded history's first remotely controlled vehicle – a kite with a piece of string as a down-link to the controller on the ground.

During the Civil War tethered balloons were used as photo platforms to seek out information on enemy positions but not until the first World War were attempts made to design and launch an unmanned powered vehicle.

The first such government contract was issued to Charles F. Kettering of General Motors a scant nine years after the Wright brothers were awarded a contract for the first manned aircraft by the Army Signal Corps.

The Kettering Aerial Torpedo of 1918 could carry 180 pounds of explosives, cruise at 55 miles per hour with a range of 40 miles. It was actually the first of off-set missiles to be followed by sophisticated 'smart' bombs of WWII. The Kettering 'Bug,' as it was referred to by the Signal Corps, was guided to its target by pre-set controls and had jettisonable biplane wings. It foreshadowed later Army attempts at over-the-hill mini-drones.

Next after Kettering came Elmer Sperry who successfully flew an unmanned N9J Navy aeroplane using one of his gyroscopes as its control unit.

In 1924, after the first World War, the Army Air Corps Engineering Division initiated a program to develop radio controls for unmanned aircraft and in 1928 attempts were made to adapt a commercial Curtiss Robin airplane to carry bombs while under control by radio.

By late 1940, the Materiel Division began a greatly expanded program to develop a variety of remotely controlled target planes and also offensive guided missiles.

Among these special weapons were the GB-1 glide bomb and the VB-1 'Azon,' followed by the 'Razon' and 'Tarzon.'

The value of the Azon was best demonstrated in Italy against railroad and bridges and in the Burma theater. In September 1944 combat missions were flown out of England with the GB-4, a television-radio controlled glide bomb.

Following World War II emphasis was placed on using war-weary or obsolete manned aircraft as drone targets. However, it was inherent that unmanned vehicles would be smaller, less costly and generally more maneuverable than converted manned aircraft.

THE GUIDED MISSILES SECTION of the Air Force was formed in 1946. From this Section came the first Pilotless Aircraft Branch, the granddaddy Project Office for later target and reconnaissance drone configurations.

The end of the war heralded a sharp decline in the military requirement for manned aircraft. Participation in staying solvent in the aircraft industry is similar to riding a roller coaster – peaks of prosperity, and depths of bare survival.

A turning point toward the possibility of staying alive in a new product area came in late 1946. Proposals were solicited by the Air Force Target Branch for two types; the **Q-1**, a 350 mph target, and the **Q-2**, a 600 mph target. Characteristics were also drafted for a supersonic target.

The contract for the Q-1 was awarded to Radioplane Company whose early designs were based on the powered radio-controlled model planes pioneered by motion picture actor Reginald Denny, a model airplane buff.

According to a 1948 Pilotless Aircraft Branch document the Q-2 requirement called for "an aerial target for realistic anti-aircraft gunnery practice against aircraft with speeds approaching that of sound; and for air-to-air gunnery practice."

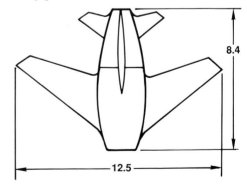

1948 AIR MATERIEL COMMAND preliminary sketch of 'Aerial Target - Type Q-2.' No contractor had yet been selected.

Thirty-one companies threw their hats in the Q-2 ring. Eighteen proposals were considered, and in August of 1948, the Ryan Aeronautical Company was awarded the first contract for a subsonic jet-propelled unmanned aircraft.

This historical transition from props to jets marked the beginning of an era wherein computerized aircraft would replace the pilot and the start of a race within the industry to participate in the potential aerial target and reconnaissance market.

THE FIREBEES

WEBSTER DEFINES A DRONE as the male of a bee (as the honey-bee) that has no sting and gathers no honey. The designation of the initial Q-2 target drones as '**Firebees**' combined the prefix 'Fire' (from the Ryan FR-l 'Fireball' Navy fighter) with the 'bee' drone insect description. Most of the Ryan Firebees had no sting (later some were armed), but their successors did gather quite a bit of 'honey' in the way of intelligence-gathering during reconnaissance missions in Vietnam.

For 40 years the basic Firebee, through constant upgrading of its control systems and performance, has been the choice of the services as its most-used jet target for ground-to-air and air-to-air crew training and weapons evaluation. In its target training and reconnaissance versions, over 7,000 Ryan unmanned air vehicles (UAVs) have been produced.

Training targets and reconnaissance birds are air-launched from the wings of a mother plane in flight, or ground-launched from a stationary or mobile installation.

For re-use, they are recoverable by parachute and retrieved on landing, or are snatched in mid-air by a MARS (mid-air retrieval) helicopter.

SHORT, STUBBY AND RATHER pugnacious in appearance, the Q-2 when fitted with a Continental J-69 jet engine, met the specifications for a target that would fly at 521 knots and attain a service ceiling of 40,000 feet.

By the spring of 1951, the first powered free-flight of the XQ-2 was accomplished, and 32 of the birds were ordered for production. This heralded the beginning of a tri-service relationship with the Firebee that has survived four decades of providing a suitable target to meet the many and varied defense needs of the Army, Navy and Air Force.

The later production targets were known as **Q-2A**s by the Air Force, the **KDA-l** and **KDA-4** by the Navy, and the **XM-21** by the Army. By 1954, substantial quantities had been ordered by the three services. In 1956, Canada joined with the RCAF ordering 30 **KDA-4s**. Came 1958 and nearly 1300 Q-2 type birds had been ordered by the services, and pilots, gunners and missilemen were finding the small jet-plane-without-a-man-in-it to be realistic threat simulation.

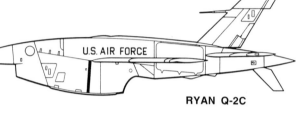

RYAN Q-2C

In 1959 an updated Firebee **Model Q-2C** went into production with new **BQM-34A** nomenclature for the Air Force and Navy and **MQM-34D** for the Army.

1958 through 1971 were the early golden years for the Q-2 Firebee series of drones. Approximately 4068 saw service at military test ranges throughout the world. The Navy was a prime customer, using 2251 birds, with the Air Force flying 1305 and the Army shooting against 512.

SINCE THE MID-50s, as the state-of-the-art in pilotless aircraft expanded, so too has the terminology used to describe this broad field. The new terms were needed to cover the growing spectrum of high-tech machines that incorporate features and capabilities never conceived in the early stages of 'drones.'

TRI-SERVICE FIREBEES of early '50s *included from front: U.S. Air Force Q-2A, Navy KDA-1, and Army XM-21. Royal Canadian Air Force Firebee, rear.*

Ryan Aeronautical Library

16

Perhaps robot aircraft is appropriate. So too are radio controlled aircraft; remotely piloted vehicles (RPVs); cruise missiles; special purpose aircraft (SPAs); the newest pilotless target aircraft (PTAs) and many others.

One who writes knowingly on the subject is Bob McVicker:

"With powerful onboard computers, precise inertial navigation systems and accurate position updates from satellites, they can operate autonomously for long periods at great distances from their bases, carrying out complex missions that are too dangerous, expensive or tedious for manned aircraft.

"Whatever their names, the goal has been the same: to remove the human pilot from the cockpit. A primary benefit has always been to eliminate the risk of human life on hazardous missions."

So today the entire field of such pilotless aircraft is encompassed in the new term 'UNMANNED AERIAL VEHICLES.' That is, UAVs.

And, what kinds of missions are possible with UAVs?

Reconnaissance; surveillance; target acquisition; target spotting; electronic and meteorological collection; bomb damage assessment; air sampling, disruption and deception with decoys; electronic countermeasures; high-resolution photography; and many more new capabilities.

NARPV and AUVS

JUST AS CAPABILITIES of UAVs have changed, so too have government procurement procedures. For many years each of the services, Army, Navy and Air Force, established its own requirements, conducted competitions and awarded its own contracts.

However, in some instances where mission requirements and commonality of support equipment were the same or similar, the services put out joint requests for proposals and awarded contracts under joint funding.

Business practices in the aerospace industry, and its relations with government agencies have also changed. One significant development resulted from a Vietnam-era noontime meeting in Dayton, Ohio.

Because of regular contacts and mutual interests, Robert T. Boone, TRA representative at Dayton, and William Mallios, senior civilian engineer in the reconnaissance and electronic warfare directorate at Wright-Patterson AFB, often discussed common problems over a business lunch.

Reminded of the occasion, which took place in late 1971, Bobby Boone recalls that "we felt the 'Old Crows' had a forum for electronic warfare and perhaps we should have one, too, for remotely piloted vehicles.

"The acronym RPVs was just coming into usage in connection with reconnaissance in Vietnam; before that 'drone' was the applicable term.

"We found there had been a number of people interested in a formal association but support for the idea was disjointed and from various places. We felt it would be useful to have a regularly published organ and an association for RPVs – a society where we could come together and share our goals and perhaps get some support, especially in the Congressional area so the technical effort could get some funding.

"We talked about it for a couple of months, and I guess it got bandied about. Then one day I got a call from Col. Earl Babcock, Deputy Director, Special Programs Office (SPO). He said, 'I understand you and Mallios are trying to get an organization going.'"

Babcock indicated there were others who were interested and suggested a larger group get together, pool their resources and interests and get going.

"The initial group," Boone recalls, "consisted of about eight of us and included, beside Mallios, Col. Babcock and myself, Col. Harold (Red) Smith, an Air Force expediter from the 147 recce program. Others were Charles B. Bagwell of Lear-Seigler and representatives of Northrop and other contractors.

"We became the nucleus of the new organization and we adopted the name National Association for Remotely Piloted Vehicles (NARPV) in December 1972. We had monthly meetings and tried to get a journal off the ground.

"By then cruise missiles were getting up a head of steam and some felt that our focus, just on RPVs, was too narrow.

"The RPV adherents were a hard-charging dynamic group of opportunists who could turn a critical need in the military theater into a new RPV overnight, as was done in Vietnam.

"But, it was soon obvious that a successor organization with a broader interest in all kinds of unmanned vehicles would eventually take over."

Additional background comes from N.C. (Dutch) Heilman who teamed with Ted Knache to put out the NARPV newsletter and Journal.

Early on, says Heilman, "The small band of enthusiasts recruited more members. Adopting the Kettering Bug as a symbol, the original Dayton chapter grew to 300 members."

"One of our projects was to hold the first National Symposium in Dayton. Soon new chapters were started in other cities with major interest in RPVs – San Diego, Salt Lake City, and Washington, D.C.

"It became clear that RPVs were only one facet of the unmanned vehicle world. There were similar vehicles operating on land, sea and in space. To encompass all, NARPV became the broader-based Association for Unmanned Vehicle Systems (AUVS) in October 1978, with headquarters in Washington."

JOINT PROJECT OFFICE

IN THE SPRING of 1988, the Defense Department set up a steering committee to oversee remotely piloted vehicle programs within the military services.

A year earlier Congress had ordered a consolidation of military RPV programs to eliminate redundancy, and froze funding pending a master plan to be submitted for approval.

Under the joint service master plan, coordination would be through the Joint Cruise Missile Office headed by Rear Adm. William Bowes.

"The congressional action," Bowes said, "while appearing severe initially, was in retrospect extremely positive. It has resulted in a profound and highly constructive change for the acquisition of UAV systems for the U.S. armed forces."

The Joint Project Office for Unmanned Aerial Vehicles, administratively under the Naval Air Systems Command, became operational in June 1988. Instead of funding individual services' projects, UAV funding is now provided at the Office of Secretary of Defense level with centralized management which coordinates operational and procurement among the services.

As Program Executive Officer, Rear Adm. George F.A. Wagner assumed responsibilities in February 1991 for the Cruise Missiles Project and Unmanned Aerial Vehicles Joint Project.

Admiral Wagner's prior assignments are particularly relative to his new command. He has been a Weapon Systems Acquisition Manager, and was Program Manager of the Navy Ship-Launched Tomahawk Cruise Missile Program.

In the manpower, personnel and training area he was tasked to improve the effectiveness and efficiency of the defense acquisition process.

EVEN AS THE CONGRESS and military services got their act together, so too have those in industry who design and build the vehicles. They have joined with the users of UAVs in common interest through participation in the non-profit Association for Unmanned Vehicle Systems.

This new melding of government, industry and academe "seeks to serve as an educational service to the UAV community and the general public. . . to broaden understanding of the potential benefits of unmanned vehicles."

While the Association covers the entire spectrum of unmanned vehicle systems – air, land, sea and space – this book is devoted solely to Unmanned Aerial Vehicles (UAVs) with emphasis on those pioneered by Teledyne Ryan Aeronautical.

GOING SUPERSONIC

FIREBEE II

EVEN AS ATTENTION in the late '60s-early '70s was focused on the accomplishment of RPVs in their role on U.S. reconnaissance missions in Vietnam and in the Middle East by the Israelis, new requirements, this time for supersonic RPVs for target training, were coming to the fore.

The two World Wars sped the advancement of aviation technology, but it wasn't until Chuck Yeager broke the sound barrier in 1947 that the goal of supersonic flight was finally attainable.

As late as 1965, target manufacturers were able to match the speed of jet fighters with their subsonic RPVs, but manned aircraft technology was soon pushed to the point where supersonic speeds were demanded for all fighter planes. While the subsonic Firebee I was still an elusive challenge for the then current crop of fighters, the time had come to test fighters' combat readiness with a target which could match or exceed supersonic fighter capabilities.

Meeting that need, Ryan Aeronautical in 1965 had been awarded a contract for development of its **Model 166** training target.

REAR ADMIRAL JAMES H. SMITH had been a Naval Aviator since 1939, and was well-qualified to represent the Naval Air Systems Command at formal roll-out ceremonies for TRA's Supersonic **'Firebee II'** at Pt. Mugu in March 1968.

He told the assembled audience, "We have developed elaborate training programs to give our fighter pilots the advantage of knowing what surprises they might expect on the line of action. We have gone to great lengths to simulate reality in this training effort. The development of a system like the Firebee II is part and parcel of this training program."

This 'formal' appearance of the **XBQM-34E** marked more than five years of preliminary studies, prototype development, manufacturing and flight test. While five years from concept to actual flight was an excellent accomplishment for such an untried and sophisticated system, the cooperation between contractor and the Navy assured attainment of required goals.

Many factors are involved in achieving a successful evolution in aeronautical advancement. The requirement for something better is always there, but without funding, that need remains a neglected stepchild.

Every successful program needs another element and that factor is the human equation: a pusher, a doer who knows the intricacies of the military procurement system. And on the other side of the fence is the engineer, the believer and the organizer who can marshall the forces within the technical civilian complex to come up with the concept to meet the requirement.

Since the early multi-service contracts for Firebee I, Bill G. Dowell, of the Navy's Bureau of Weapons, had been an enthusiastic supporter of drones and the **BQM-34A**s in particular.

As one of the top civil service employees assigned to the Naval drone programs, he was not only familiar with the birds themselves, but unusually aware of the funding situation in BuWeps. Strictly impartial, Bill dealt firmly with all RPV companies and guarded the Navy purse strings as though they were his own. In addition to this background, Bill had the foresight to see the eventual need for a faster, higher flying target for the Navy's new breed of fighters.

Dowell's contractor counterpart was one of Ryan's aeronautical engineering believers, Sam Sevelson. Born and raised in South Africa, Sam served a hitch in their Air Force

AGAINST FIREBEE II MOCK-UP, engineers *Ron Reasoner, Sam Sevelson and Carroll Berner plan the new target's future.*

as a pilot, and migrated to the States to get his degree in engineering in Texas and California.

In late 1962, Dowell made several exploratory visits to Ryan and other target manufacturers, looking into the possibilities and problems involved in getting a supersonic target. The Dowell/Sevelson merger of brains and drive was the beginning of a long association that ultimately resulted in the **Model 166 Firebee II.**

The third vital part of the formula — funding — was not available in sufficient amounts for an R&D program. However, Ryan recognized the eventual need for a supersonic bird, and gambled on the team of Sevelson, chief engineer Erich Oemcke and Ron Reasoner, chief project engineer, to combine their talents in pushing the concept.

Dowell managed to scrape up forty thousand bucks as seed money to help defray some of the initial costs. Armed with a comprehensive feasibility study, and subtly directed through the Navy procurement channels by Dowell, Sevelson, Reasoner and the venerable Charlie Smith, the company's Washington rep, spent several months briefing top Navy brass on the virtues of the growth version of the BQM-34A, a '**Super Q-2C** Firebee.'

By midyear 1965 their efforts were successful.

BUWEPS ISSUED A CONTRACT to determine if the subsonic Firebee could be upgraded to meet a Mach 1.5 speed and altitudes in excess of 60,000 feet. Also required,

5G maneuvers up to 20,000 feet and 2 to 3G maneuvers at higher altitudes. The concept of a growth version allowed the Bureau to expedite procurement, but upgrading the chubby BQM-34**A** into the sleek BQM-34**E** took more than a little doing.

The precedent for this type of procurement had been established with the 'Big Safari' programs for the RPVs supplied to the Air Force in the Vietnam conflict.

Two and a half years elapsed between the signing of the initial contract and the maiden flight of the new bird. Transforming a buxom grandmother into a scintillating ballerina was no simple task, but the team of Oemcke, Reasoner, Sevelson — all under the guidance of Bob Schwanhausser — not only recognized a shapely form, but also how to make one.

One of the pressing problems facing the engineers at the start of the design phase was the reduction in size of the fuselage profiles. While the CAE J69-T-29 engine nestled comfortably in the tummy of the BQM-34**A**, there was no way it would fit into the slim profile of the 'Super Q-2C.' CAE responded to the challenge by remounting the accessories in a compact design closer to the main body and reconfiguring the exhaust system for a cleaner, more streamlined exit aft.

Redesignated as the YJ69-T-6, its power was upgraded to 1850 lbs. thrust (150 more than the T-29) and gave the new bird enough boost to make the transition from subsonic to supersonic speeds. With her girdle now fully tightened by Continental, the fat lady appeared to be a successful graduate of the weight-watcher program.

In addition to the frontal area being reduced to a 25-inch outside diameter, the compressor was redesigned to uprate the thrust, and material changes in the radial compressor permitted the desired supersonic operations at sea level.

Stagnation temperatures in the high supersonic regimes required some pioneering effort in the engineering department. The final con-

CUT-AWAY PROFILE OF Supersonic Firebee II *is used in accurately displaying placement of all equipment.*

21

figuration for the wings and control surfaces was a thin stainless steel covering wrapped around a bonded aluminum honeycomb core. Titanium was used to protect the leading edges.

Compared with Firebee I's 23-foot length, the Firebee II supersonic version stretched out to 28 feet, while wing span was clipped from 13 feet to 9 feet.

The problem of combining sub-with-supersonic speed and enough endurance aloft to meet mission requirements led to the use of a jettisonable 58-gallon belly slipper fuel tank. With the tank attached, the frontal cross section of the bird was almost identical with the standard BQM-34A, and performance throughout all ranges was much the same. It was only after the tank was dropped that the new bird could achieve supersonic speeds.

Design and construction of the belly tank caused some unusual situations. Sevelson noted, "The darn thing was built to withstand the anticipated high G forces on ground launch, but the final product looked like it came out of the Ryan Iron Works rather than an aircraft factory. One Navy bean counter said that the tank cost as much as a Cadillac, and with Caddies selling at around six thousand bucks at that time, he was right."

Following climb after launch, and preparing for the transonic sprint, the command could be given to jettison the tank which would splash down in the open seas with the potential of becoming a navigational hazard to small boats. Early on, one enterprising skipper salvaged one off the coast of Point Loma and returned it to North Island off San Diego. The potential hazard of a floating fuel tank was overcome with the installation of a corrosive magnesium plug which dissolved within minutes after contact with saltwater, giving Davy Jones another addition to his underwater aeronautical museum.

In addition to having a supersonic capability, BuAer insisted that the new bird retain all of the capabilities of the BQM-34A. Scoring, active and passive radar augmenters, ECM, infrared flares, and low altitude systems were to be available. In common with the subsonic bird, either ground launch or air launch from the DP-2E or the C-130 should be incorporated, as well as water recovery or MARS system. These criteria were all met, and compatibility with all Firebee I ground check-out equipment, direct and remote control stations were proven.

Sevelson remembers, "The initial contract was for four flight test vehicles and one static test vehicle. It included an option for ten 'prototypes,' but the money for these ten birds came later, increasing the total value of the

U.S. Navy

SUPERSONIC SURFBOARD? No! *Test engineer Walt Hamilton had simply gone overboard to free a plastic cover.*

contract to a figure which was a bargain if you consider we were designing, building and flying four test vehicles.

"We were still getting some flak from several naval procurement types, claiming we were actually building an entirely new target. My answer to that charge was 'not so — we merely changed the loft-lines.' We were also said to be making a Taj Mahal out of an outhouse, but cooler heads prevailed."

THE FIRST OF THE TEST BIRDS was delivered to the Navy's Missile Test Center at Pt. Mugu in late 1967, and the maiden flight was successfully made in January 1968.

On June 10 the target made its first truly supersonic dash, hitting Mach 1.3 at 36,000 feet altitude. Recovery was commanded and the bird descended to open ocean 300 yards from a Navy retrieval boat acting as safety monitor in the event of an emergency. And an emergency *did* occur.

The parachute riser plastic cover failed to detach and was still housing the retrieval strap to which the incoming helicopter was to attach. Walt Hamilton, then one of TRA's flight test engineers aboard the retrieval boat, didn't hesitate to put on his wet suit, mask and fins and swim to the Firebee. Using a Navy issue survival knife strapped to his leg, he pried the cover loose, freeing the riser. The helo made contact, and the target was returned undamaged to the base.

Walt felt that John Bunganich, the company base manager, went a bit overboard in reporting the incident. "Without regard for his personal safety, Hamilton acted promptly to save the situation that could have delayed the flight test program. In doing this he exposed himself to the ever-present hazard of sharks in the waters off Pt. Mugu."

Hamilton figured it was all in a day's work, not realizing that his actions added another incident to the TRA can-do legend.

A total of 23 proving flights were made in the following fifteen months. Twenty were launched from an old DP-2E Neptune, and three from the Mugu ground launch pad. Early concern over the structural capability of the slimmer bird to take the ground launch G loads proved groundless. Subsonic and supersonic requirements were all met or exceeded, with speed as high as Mach 1.68 and 63,000 feet altitude.

Satisfied with results of early test and evaluation flights, with the original XBQM-34**E** birds, major production orders were placed by the Navy and Air Force in 1969 and again in 1971.

By early June of 1971, the Firebee II reached the homestretch in a three-and-a-half-year flight test and evaluation program at Pt. Mugu. In addition to 25 successful ground launches, the prototype birds logged 51 recorded flights including 28 in the supersonic dash mode averaging 25 minutes each. All of the flight tests and developmental programs were conducted by the Navy's Threat Simulation Department.

GROUND LAUNCH with jettisonable fuel tank *attached is validated in initial tests from Pt. Mugu pad.*

BIRD AWAY! **As BQM-34E Firebee II** *is air launched from Navy DP-2E on Atlantic Fleet Weapons Range, Puerto Rico.*

GROUND LAUNCH

Teledyne Ryan Aeronautical

Dave Gossett

AIR LAUNCH

A PAIR OF WINNERS. Navy fighter pilot *surveys the new supersonic threat simulator and ponders the outcome of his first encounter.*

A year after the first supersonic flight, Lt. Col. Charles L. Zangus, executive officer of the Marine Air Detachment at Pt. Mugu, scored the first weapons air-to-air kill of Firebee II, firing a Sparrow missile in a head-on shot. The intercept occurred at 44,500 feet with the prototype Firebee II cruising at Mach 1.65 (1090 mph).

The roll-out of supersonic production model number 0001 on July 21, 1971 had all the hoopla and pageantry of a regal coronation, Vice Adm. Thomas J. Walker of ComNavAir-Pac was the ranking Naval officer, and Frank Gard Jameson, President of TRA presided as the needle-nosed bird burst from the hangar through a huge paper bulls-eye. TV and press covered the event and hundreds of employees enjoyed the break in their routine to witness the first of their new production babies christened into the Naval inventory.

Final assembly inspection of the first production vehicle began two weeks before formal delivery. It was conducted by contractor personnel and a team of technical representatives from nine military agencies.

THE NAVY'S NEW SUPERSONIC target went operational on March 29, 1972, at the Atlantic Fleet Weapons Range in Puerto Rico. This initial flight tested the USS Wainwright's Terrier missile defense system against simulated enemy aircraft or anti-ship missiles, closing on the ship from 50,000 feet, accelerating to Mach 1.52 during penetration of the ship's defenses. Onboard sensing devices scored a successful near-miss without destroying the target.

The flight provided two presentations, and was controlled from the defense center by Dick Manceau, TRA's Base Manager. The first run was a subsonic tracking run to allow the guided missile destroyer Wainwright to look at the target and the controllers to get the feel of maneuvering the new target. The second run was supersonic.

Launch was from an aging DP-2E, and sea pickup by helicopter following chute deployment indicated no damage to the bird. Captain W.A. Mackey, Rosie Roads Range

UNVEILING OF FIRST FIREBEE II production vehicle *features presence of VADM T.J. Walker, ComNavAirPac; program manager Bill Dowell, CDR Jim Brady, and Frank Jameson, TRA President.*

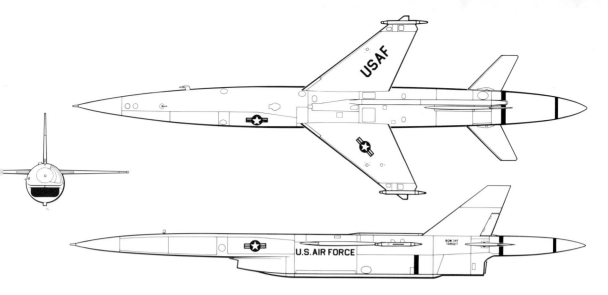

commander, called the air defense missile exercise a "significant success." Firebee II had officially become operational and was ready to meet the Navy's supersonic target requirements for years to come.

Thirteen months following the first operational flight, Firebee II joined her predecessor in one of the largest operations held at Rosie Roads. U.S. Atlantic Fleet as well as British and West German naval units combined to measure the defense readiness of the units engaged.

Seven supersonic Firebee IIs — the highest number in a single operation — were flown, simulating air superiority threats against surface as well as air units.

Twenty-five Firebee Is were also flown during day and night missions in 34 hours on June 5-6 in one of the most demanding support operations on record for Mike Savino's 20 man team of contractor specialists.

Launch operations were conducted from Anagada Island, and also from a launch boat some 80 miles out at sea for the night firings, and from DP-2E Neptune aircraft for the daytime operations.

The U.S. aircraft carrier 'Franklin D. Roosevelt' and British carrier HMS 'Ark Royal' pitted their squadrons against both classes of targets, while seven U.S. and a West German missile ship, the 'Rommel' tested their surface-to-air capabilities.

For some of the units engaged in this exercise, the results were to have a far-reaching effect. The operation was a critical testing ground for operations in the Mediterranean Sea in the years ahead.

Supersonic Firebee IIs continued to be available for operation at Pt. Mugu, and a small inventory was also maintained at Roosevelt Roads, into the late '80s.

FROM THE TIME of the original Navy contract in 1965, the U.S. Air Force constantly monitored the progress of the supersonic bird through its engineering and flight test phases. Confident that the program was to become a success, they joined the Navy in placing the initial production order for Firebee II targets in 1969. A small order for engineering changes followed in 1970, and in May, 1971 the initial delivery of the **BQM-34F** (Air Force designation) was made.

Interservice cooperation was demonstrated at Pt. Mugu in August, 1972 when the first official Air Force supersonic target flight was conducted. The bird was air-launched from a DC-130 at 10,000 feet, climbed to a cruise altitude of 50,000 feet during the one hour and fifteen minute flight. The flight was termed flawless, and the stage was set for the formal Air Force roll-out at the Air Defense Weapons Center, Tyndall AFB, Florida, during the William Tell Weapons shoot in September.

Preceding the William Tell events, the 6514th Test Squadron at Edwards AFB conducted helicopter mid-air retrieval tests (MARS), snatching the supersonic targets while they were still descending to ground by parachute.

The MARS system was used to cut maintenance time, lengthen life span and reduce turnaround time in restoring Firebee IIs to flight status.

During tests at Tyndall in summer 1973, Lt. Col. Harold D. Wilson, commander of the 4750th Test Squadron, became the first Air Force pilot to 'kill' a Firebee II. Wilson used a radar-type missile launched at 40,000 feet from his F-106 Delta Dart.

With the Firebee II flying at Mach 1.1, the interceptor's pilot had to close on the target at Mach 1.3. The AIM 4-F missile struck the drone in the left wing and fuselage area. "It is always a big thrill," the Colonel said, "to destroy a target in the air. It still had a big load of fuel so it was a quite spectacular hit."

Designed from the start, under a growth-version concept as a match for air-superiority fighters, Firebee IIs tangled with Air Force interceptor pilots in the 1976 William Tell competition, flying as targets for F-101s, F-106s and F-4s firing Falcon, Sparrow and Sidewinder missiles. All supersonic missions against BQM-34Fs were ground-launched and flown at altitudes of 50,000 feet during missile firings.

Firebee IIs also flew 16 missions in support of William Tell 1978. It wasn't until 11 years later that the last Firebee II was expended at Tyndall.

During the flight test program, Firebee IIs had been used in July 1972 in support of Army Hawk ground-to-air missile firings at White Sands Missile Range in New Mexico.

Later, the Army purchased several BQM-34E/F supersonic targets. But, having been designed for 'softer' water recoveries (Navy) or mid-air retrieval (Air Force), the Army found that impact with solid ground could have an adverse effect on the supersonic target's structure. Without MARS available, the Army's birds ended up with the Air Force at Hill AFB in Utah.

With the subsonic Firebee I having led the way on many of the product improvements, it was in the spring of 1976 that the improved RALACS (Radar Altimeter Low Altitude Control System) was incorporated into the supersonic version. Instrumented with the improved system, the Firebee II was ground launched at the Pacific Missile Test Center and climbed to 4,000 feet as part of a scheduled target presentation.

Low level flight was then engaged and the bird descended to 700 feet in subsonic speed mode to jettison the belly tank and then translated into the supersonic mode at 97% power and at Mach 1.1 speed.

It flew 82 seconds at an altitude below 100 feet, dropping to 50 feet for the final 20 second period of the flight operation. No missiles were fired because a surface bogey was picked up on the radar, but the flight was a forerunner of more development work to be accomplished on 'sea-skimmer' concepts.

Follow-on Firebee II contracts through 1975 resulted in a total of nearly 300 of the supersonic birds being built for the Navy and Air Force.

When the initial contract was let in 1965, Bill Dowell optimistically predicted, "This is the future. When this thing goes into production and is operational, we will have a single target to fit all future requirements." However, he was wrong in predicting the demise of the subsonic BQM-34A.

The military requirement for the supersonic configuration was considerably less than the need for the subsonic target. Money for research into higher and faster concepts for targets seriously limited the production budgets, and on completion for Firebee II orders for both services, further production was discontinued.

However, twenty-five years later, the matronly and buxom Firebee I grandmother is still providing the requirements for all three services for a reliable, subsonic target. And — the sleek young ballerina has retired to savor her scrapbook of accomplishments in the supersonic history of UAVs.

QUADRUPLE THREAT; loaded for bear! *A pair of supersonic Firebee IIs ready for launch from right wing of DC-130. A similar pair of subsonic Firebee Is are slung beneath left wing.*

U.S. Navy

Flying **HIGHER**
and **LONGER**

ALAMOGORDO DAILY NEWS, AUG. 5, 1969

Secret Drone Lands Safely And Away From Prying Eyes

LOS ALAMOS (AP)—Imagine the consternation of Air Force officials Monday when their secret unmanned aircraft began descending by parachute in northern New Mexico—150 miles from the security of White Sands Missile Range.

They feared the prying eyes of nosy civilians might reveal some of the strange bird's tightly guarded secrets.

But Lady Luck smiled on the Air Force.

Dangling under an orange and white parachute, the aircraft came to a gentle landing inside a Los Alamos Scientific Laboratory experimental area. Fences kept the public at bay. The only prying eyes were those of security conscious Los Alamos employes, who handle daily the secrets of the nation's nuclear weapons programs.

An Air Force spokesman said the drone was undergoing test over White Sands Missile P and off-range launch si

The military woul' the drone was gro' or if it was con' craft. Th' clined' dr

The area manager for the Atomic Energy Commission in Los Alamos, H. Jack Blackwell, said he watched the drone land from his office window.

Another witness said the drone fuselage was about 20 feet long and the wing span was about 40 feet.

Laboratory employes quickly covered the aircraft and awaited its removal by the Air Force.

COMPASS ARROW

NOT ALL GOVERNMENT contracts, re-gardless of their basic merit — let alone un-solicited proposals made to domestic and foreign military services — end up as success-ful projects. Timing may be bad; the political climate may have changed; new personnel may not continue the programs initiated by their predecessors; initial mission require-ments may no longer be valid.

Nonetheless, it is helpful to recall some in-complete ventures because their technical ap-proaches and operational concepts were unique. Compass Arrow, as well as Compass Cope, its successor, were such ventures.

'Compass Arrow' was born 'Firefly' and was assigned company **Model 154** and military designation **AQM-91A**. By whatever nomen-clature it was an innovative, secret program which parachuted into the spotlight by acci-dent.

SECRET SOMETHING

THE HEADLINE in the Albuquerque Jour-nal of August 6, 1969 read **"Secret Some-thing Falls to Earth."**

Written by Bill Stockton of Associated Press, the article described "the emergency descent by parachute of a super secret un-manned aircraft in full view of Los Alamos [New Mexico] residents" which "ripped some security wraps off 'Firefly,' an Air Force proj-ect hidden at White Sands Missile Range for two years."

For many years Ryan Firebee target drones used in air-to-air and ground-to-air missile training had operated relatively openly in the White Sands area. After all, the Firebee was an unclassified pilotless aircraft not subject to the more stringent security regulations. But obviously the strange craft which descended on Los Alamos was a bird of a different feather.

Continuing, Stockton wrote —

Dangling from a bright orange and white parachute, the innocent-looking 20-foot long aircraft settled gently onto an asphalt road Monday, 60 feet from a building but safe from the public be-hind the fences of a security-conscious Los Alamos Scientific Laboratory tech-nical area.

Reverberations from the thud of the graceful bird's landing on the northern New Mexico mountain plateau, 150 miles from White Sands, was felt all the way to Washington.

Air Force officials then confirmed for the first time that a project known as Firefly was under way, cloaked in the secrecy of the vast desert test range and nearby Holloman Air Force Base.

LANDING UNDAMAGED on its inflatable 'cushions,' *the strange 'Firefly' drone is safe but the secret is out.*

Ryan Aeronautical Library

The Air Force spokesman said Tuesday that Monday's errant aircraft was 'relatively high altitude test of an Air Force target drone launched over the White Sands Missile Range.

'Due to a slight malfunction. . . the drone was brought down on a government reservation other than the planned recovery area.'

Even with the acknowledgement that there is a Firefly, the Air Force declined to comment on other unanswered questions connected with it.

If Firefly is simply a high altitude target drone for testing [the accuracy of] missile systems [fired against it], the reason behind the strict Firefly or drone aircraft secrecy lid remains a mystery.

[The security blanket was still covering another Firefly vehicle, the **Ryan 147A model** which had conducted the original drone reconnaissance spy plane tests in 1962]

The Firefly had landed barely inside the security fence at Los Alamos, just across a narrow canyon from a residential area. Its descent had been witnessed by thousands of residents. One was H. Jack Blackwell, area manager for the Atomic Energy Commission who watched from his office window. "It came in nice and easy, about noon," he said.

Other witnesses said the drone fuselage was about 20 feet long and the wing span about 40 feet. They reported it kind of slid in, one wing slipping in under the road guard rail, but not even scratching the asphalt.

BEFORE SECURITY PEOPLE were able to partially cover the drone a number of photographs had been taken. Two of them appeared in the Los Alamos Monitor.

How anxious the government was that the pictures not be printed was indicated in a wire story filed out of Los Alamos by United Press International:

The owner of a Los Alamos newspaper said Tuesday he resisted government efforts to suppress publication of two photos showing a pilotless high-altitude Air Force plane that floated to earth here Monday.

Markly McMahon, editor and publisher of the Los Alamos Monitor, said the photos appeared in his newspaper Tuesday afternoon despite requests by several government agencies to stop the publication on 'security' grounds.

At the same time, Darrel Burns, owner of radio station KRSN here, sent

a letter asking U.S. Sen. Clinton P. Anderson, D-N.M., to look into the handling of the plane incident.

Burns said he told the senator he was "amazed that it took four hours and 15 minutes" after the landing for the public information office at White Sands Missile Range to issue a statement.

When a news release was finally made, Burns said, it "was inaccurate and irrelevant to the people of Los Alamos. When something happens you just shouldn't wait four hours to be able to tell people about it.

"If you want to say it is classified and never happened, that's alright with me," Burns said. "but don't wait four hours."

"It is not every day that you see a plane coming down on a big orange and white double parachute with a C-130 Air Force plane circling over the town," McMahon said. McMahon's wife, Jean, said she saw the craft floating down from nine miles outside of Los Alamos.

McMahon said a WSMR official and the U.S. Air Force "insisted we should not run the pictures."

Judging what went on at Los Alamos it was obvious the Air Force security and public information people had a hot potato on their hands.

It was not the first time they had been burned by the Firefly.

IN FEBRUARY 1967 a Boston brokerage firm put out a confidential newsletter to institutional customers with "Notes on. . . the Air Force's 'Dark Eagle' Program." It went on to describe participation of other companies "with Ryan Aeronautical in a multi-million dollar Air Force effort to develop a high-altitude, long-range unmanned reconnaissance aircraft system. The program is believed to be known as Dark Eagle."

Fortunately the confidential newsletter attracted little outside attention and the press never made any inquiry of Ryan regarding the Dark Eagle at that time. Barry Miller, a knowledgeable technical writer with Aviation Week magazine, was identified as the writer of the brokerage firm's newsletter.

NEARLY A YEAR LAPSED before the Army's White Sands Missile Range inadvertently mentioned the Firefly project in a January 1968 wrap-up release outlining the range activities during the year ahead:

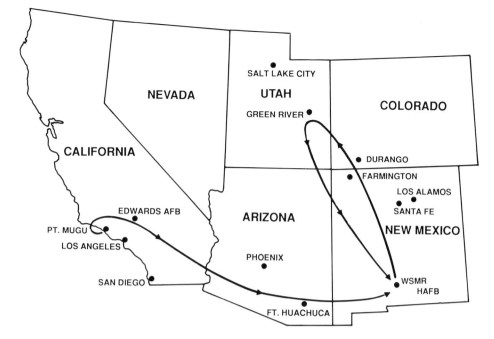

Firefly will test a special purpose vehicle. Initial flights will originate over the White Sands Missile Range with later flights originating from above the Pacific Ocean and terminating over the White Sands Missile Range.

The early flights will follow patterns from WSMR to the vicinity of Green River, Utah, and back to the range. White Sands will monitor all flights and provide range support.

At the time, newsmen asking for more information were told no more was available and were referred to Air Force officials who said only that details were classified and no information beyond that in the Army release was available.

Privately, Air Force sources labeled the release a "real booboo," and were pleased that it attracted little attention.

Then on November 22, 1968 the lead item in Aerospace Daily attracted unwanted congressional attention.

The trade publication reported that "Ryan's super-secret Firefly reconnaissance drone now is undergoing flight test at WSMR." It went on to explain that earlier in the year test of its MARS II mid-air recovery system had been conducted at the Joint Parachute Test Facility, El Centro, California.

The current series of flight tests, it said, followed the plan outlined by White Sands in its "booboo" release of January.

But what caught the attention of Rep. Morris K. Udall (D-Ariz.) was the fact that civilian authorities were not aware of the planned future launch of a Firefly above the Pacific Ocean to overfly California, Arizona and New Mexico and to be "recovered at White Sands."

The Daily noted that an instrumented range with sophisticated tracking radar and telemetry receiving equipment is maintained by the Air Force from WSMR to Green River — about 940 miles — to support advanced ballistic missile reentry vehicle tests.

It was assumed the drone would make the 1200-mile flight beginning off the Pacific Coast, then over Edwards Air Force Base to the Army's Fort Huachuca, Arizona and on to White Sands, "as this route already is an established instrumented range used by the Air Force in testing aircraft inertial guidance systems.

"Overflight of populated areas by an unmanned vehicle," the Daily said, "would require constant monitoring by radar to ensure safe operation.

"Extreme sensitivity on the part of DoD about its reconnaissance programs since the U-2 affair has caused these activities to be shrouded in complete secrecy. This would account for the fact the Air Force has not followed its traditional course and informed the public (within the bounds of normal military security) of the planned over-flights. In the past, advance coordination with state and community governments, civic agencies and the press has preceded overflights. . ."

Representative Udall reported that he had no knowledge of such a program and that DoD had not coordinated with him. "I'm quite concerned," he told the Daily. "As a private pilot I'd hate to run into one of those things." It was obvious Udall would be asking the Department of Defense for an explanation.

GHOSTING THROUGH THE SKY, the elusive 154 test vehicle *is captured in high-altitude free flight by a ground-based cine-theodolite camera.*

Teledyne Ryan Aeronautical

FROM WHITE SANDS Missile Range, the Associated Press reported several weeks later that "The Air Force is apparently departing from past policy in clamping the secrecy lid on Firefly.

"Military officials went to great lengths to work with public officials and inform residents of Utah and New Mexico about the Athena program that began in 1963 in which missiles are fired from Green River, Utah, to White Sands.

"Firefly was first thought to be a guided missile, but speculation has shifted to the possibility it is an unmanned reconnaissance aircraft to photograph enemy territory and return for a parachute recovery."

By January 30, 1969 Aerospace Daily was able to report the results Rep. Udall and Rep. Alonzo Bell (R-Calif.) had obtained from the Air Force in response to their inquiries regarding the safety of the planned California-New Mexico overflights by the secret Firefly drone.

As explained by the Air Force Legislative Liaison office, "In the case of drone tests, launch and recovery is conducted in military-controlled airspace over established testing ranges. Additionally, when such test drones traverse other than military airspace they are equipped with radar beacons to allow the FAA to follow the flight progress and provide separation from other known air traffic. The FAA is provided the predicted flight plan well in advance of such tests.

"We assure you these safety measures will be followed whether or not the flight was officially announced."

Previously, according to the staffs of the congressmen, the Air Force after first denying any knowledge of the program confirmed that there was a Firefly project, that it dealt with a drone flight test and was classified. They also referred the congressmen to Aerospace Daily for further details, adding that they "could neither confirm nor deny the accuracy of the story."

There the matter rested until the August day, about seven months later, when the Firefly floated down unannounced at Los Alamos.

But just what was the Firefly?

Firefly was Ryan's **Model 154**. It was new and far more sophisticated than any pilotless jet aircraft ever before flown.

HIT THE PANIC BUTTON

THE FOURTH PRODUCTION vehicle (P-4)in the new 154 Firefly series of Ryan reconnaissance drones was flying steadily over north-eastern New Mexico on a test flight that August day in 1969. Under control of Holloman Air Force Base, it was flying on a planned off-range mission on pattern C beyond confines of the White Sands Missile Range.

The navigational test flight of the recently developed high-altitude drone was also being monitored by the Air Force C-130 director aircraft from which the 154 had been airlaunched three hours earlier.

Suddenly the panel lights at Holloman ground control lit up signaling the fact that a control surface actuator on the drone had failed. In seconds the bird would be out of control. Nothing to do but hit the 'panic button' which would release the parachute on the 154 permitting it to descend with minium damage so that it might be recovered.

The C-130 cruising the area, got the same word and was immediately vectored by Major Chuck Smith at Holloman to the last plotted position. The 154 was still swinging beneath its 100-foot parachute canopy, descending slowly, when the C-130 arrived in the vicinity of Santa Fe, New Mexico.

Aboard the launch plane with the Air Force crew were two Ryan technicians — veteran Dale Weaver and Art Rutherford. "Everyone was looking out the windows," Rutherford recalls, "to see if they could spot the bird. We were flying over an area of deep arroyos, with a big complex of factory buildings or hangars in between.

"We spotted the bird in the chutes and spiraled down with it. About the only comment from the pilot, Major Ken Beckner [Major John Wyman, co-pilot], was "I don't like it. I don't like it," referring to where the 154 was going down. He had reason to be concerned not only because of possible damage to the drone, but also because security of the secret new reconnaissance vehicle might be compromised. The whole program was in jeopardy.

"We watched as the bird descended and came to rest, apparently undamaged, on a service road adjacent to one of the large buildings of the Atomic Energy Commission complex at Los Alamos.

"Later Dale was to learn from one of the AEC firemen that the drone was dripping fuel as it came down, drifting over the main plutonium processing plant. Had it gone down there they would have had to evacuate the area. If the bird had impacted the special

CHUTE TO THE RESCUE! Failure of a control surface **actuator** *while over New Mexico prompted the command for an unscheduled landing.*

Teledyne Ryan Aeronautical

materials building and started a fire, it would have taken days or weeks to put it out. We were just lucky.

"Bob Schwanhausser and an Air Force group were back at Holloman. Mission control gave the pilot of our C-130 information on the nearest air strip. It was right there at Los Alamos. We made a pass over the field but because it was quite small and at a relatively high altitude the pilot elected to go into the larger field at Santa Fe. There we piled out of the C-130 with our tool kits, picked up several rental cars and were on our way to Los Alamos.

"Our party consisted of Captain Stan Worth, the airborne remote control officer, two Air Force launch control operators (LCOs), two sergeants and Dale and I from Teledyne Ryan.

"On the edge of Los Alamos less than an hour later, still clad in flight suits, we stopped at the Highway Patrol office and Dale went in to get directions. He got them alright but not until the desk sergeant had kiddingly told him 'I'm not going to tell you where it is until you tell me what it is.'

"The AEC was expecting us and were very accommodating, issuing badges which gave us access to the area where our bird was parked. Their public information director was sympathetic to our problem, being aware of the sensitivity of our type of project but as Dale pointed out there just was no really good 'cover story' available. Their three-story hangar was closed — it appeared to be a highly classified area — and the AEC security people gave us a pretty good going over coming and going.

"**A** TEN-FOOT HIGH FENCE secured the perimeter about 100 yards out from the building. Closer to the hangar was a 35-foot asphalt drive with a guard rail. Our Special Purpose Aircraft (SPA) had landed right in the middle of the driveway, the right wing skidding in under the guard rail.

"By the time we arrived the news people were there, pressed up against the perimeter fence or up in trees to get a better overview. Nothing prevented them from shooting pictures of the 154 with telephoto lenses for the papers and local TV, or using their tape recorders for first hand accounts of the unscheduled landing of a 'secret something.'

"The glass windows of the drone's camera compartment were frosted over from normal condensation so no one could see into the 'scorer' compartment.

"Several of the AEC employees were knowing enough to get some tarps and partially cover the bird as they realized it was probably classified. There were a lot of grins but they were wise enough in the ways of security not to ask questions they knew we couldn't answer.

"We were at a loss to know what the hell to do with it, stuck there under the guard rail. The bird had landed during AEC's lunch hour which is apt to be pretty much an employee picnic type of occasion. They had a grandstand seat.

"One employee was a lunch hour athlete who regularly jogged around the perimeter driveway. As he finished his third lap and came around on his fourth, there was this huge bird plopped down in the driveway in front of him.

"It wasn't just the bird floating down in its chute which had attracted attention. To keep it from impacting too hard the 154 is equipped with inflatable attenuation bags. They'd been set to activate at 9,000 feet, but Los Alamos is about 7,000 feet elevation so things had started to happen at a relatively low altitude for the spectators.

"HERE'S THE BIRD FLOATING gently down beneath its 100-foot bright orange parachute, peaceful as can be, when suddenly there are big flashes and loud explosions as more gear starts to deploy as the doors covering the inflatable bags are blown off. Debris is falling and everyone has reason to believe that the bird may be ready to self-destruct.

"Dale was in contact with King One — mission control at Holloman — and they were up in arms as to what to do. The Air Force and Ryan people there put together a team and took off in a C-130 for Los Alamos. Meanwhile Kirtland Air Force Base at Albuquerque — much closer to Los Alamos — was to start a group on the way.

UNDAMAGED, WITH A PILLOW-SOFT LANDING, the Model 154 *awaits pick-up on the White Sands Missile Range.*

"We went ahead with the tools we had to remove the wing tips, get off various fairings and do as much of the disassembly work as we could preparatory to transporting P-4 back to Holloman.

"The rigging boss from a subcontractor offered assistance and we said 'fine, if you can help us lift it up.' He came back with a couple of cranes, a fork lift and other handling gear and steel saw horses. He lifted the bird out and placed it on two of the saw horses. By then it's after dark. By the time the crew from Holloman arrived we had pulled the nose and empennage off, and it was damn near ready to take apart and load on the flat bed truck for Santa Fe where the C-130 would take it on home.

"Dale had signed some type of purchase order with the rigging company and I guess it was months before the paper work between the Air Force, the Atomic Energy people and the Ryan types back in San Diego got straightened out and everyone paid for the emergency assistance they gave us.

"By the time we got back to Santa Fe we'd been working continuously for 28 hours — our preflight had been at 2:30 a.m. the previous day — and were a pretty scroungy group. When they got the disassembled 154, all the gear and crew aboard the C-130 we were badly overloaded so Dale Weaver, Harry Henninger, the Ryan maintenance supervisor from Holloman, and I volunteered to come back by commercial airline.

"Harry had a razor and we did the best we could under the circumstances but we were a sad looking contingent in tired flight suits who deplaned later that day at Alamogordo. By then every one was reading the account of our adventures on the front page in the Albuquerque Journal."

AUTOMATICALLY INFLATED at 9,000 feet, *the droopy bags guarantee a softer landing.*

THE IMPORTANCE OF QUALITY workmanship in assuring product reliability of a complex vehicle like the 154 was well demonstrated by the unscheduled Los Alamos landing.

"This particular accident," recalls engineer Larry Emison, "was traced back to a relatively small and seemingly unimportant incident.

"An actuator is the heart of the autopilot that controls the elevator. When you lose that, you have lost control of the vehicle and have to take emergency measures.

"The problem was ultimately traced to the vendor who had designed the brushes in the motor part of the actuator. His drawing specified that high temperature solder be used in attaching the wire to the brushes. Someone in his shop failed to comply. The workman either deliberately ignored the specification or simply went ahead without checking and used low temperature solder instead.

"The result was that when the part got warm in actual use during flight the wire separated because the solder got soft. End of flight!

"Here were millions of dollars and an important defense project placed in unnecessary jeopardy because of one small piece of sloppy workmanship. We could have lost the bird. We did lose the data we were after. And, I suppose, the whole project could have been cancelled if someone really wanted to make an issue of it.

"Not only that but we were grounded for weeks and weeks after that because the missile flight surveillance people on the White Sands Range asked for a full investigation. Even after that they restricted us to the range until we could conclusively demonstrate that we had fixed the problem."

MEETING THE CHALLENGE

THE PHOTO RECONNAISSANCE challenge has always boiled down to this: How to fly higher, farther and more precisely with less vulnerability to enemy weapons, and return with high-resolution photographs which yield significant intelligence.

Even as the 147 family of Ryan drones was gaining recognition and acceptance for the important new dimension they were adding to American reconnaissance capability, the search started for a still more advanced system.

The Asian land mass – Red China – represented a challenge beyond the capability of the 147 family. The Nationalist Chinese were filling some of the requirement with operation of U-2 manned vehicles out of Taiwan, but the risk of capture was ever present, and in any case the United States preferred to have its own capability.

Thus, in 1965 Ryan brought to a head discussions with the Air Force regarding this country's longer-range requirements. As the most knowledgeable drone specialist, Ryan was in a unique position to undertake development of the next generation recce drone.

Experience with the 147 series of vehicles indicated that increased altitude is the best single defense against most defensive sensors. Therefore, studies which led to the Model 154 design were centered on achieving maximum

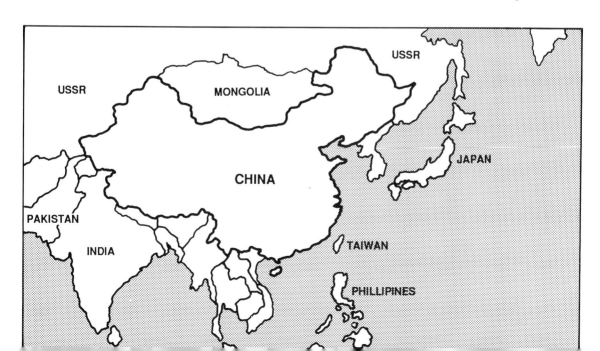

cruise altitude during the mission segment. And at this altitude, visual detection would be minimized by cruising above those levels which produce contrails. Proper shaping and arrangement of the aircraft, positioning of the engine on top of the fuselage, infrared suppression and use of radar-absorbent materials, and other techniques could be used to further reduce vulnerability. (An early example of Stealth technology.)

Although the 154 project finally went under contract in June 1966, it had been years in the making as explained by Bob Schwanhausser:

"Back in 1960 when we were talking with the Air Force about 'Red Wagon,' one of the possible configurations was a bird that looked an awful lot like the 154. But, of course, that advanced concept would have required funding for an entirely new vehicle, and you'll recall we had to settle in the early days of the program for the 'Big Safari' approach of modifying already existing drones. So we didn't go ahead with new hardware at that time.

"However, when we came back to San Diego from one of our Washington conferences, one of the designers sketched in a pair of wheels on the design we had proposed. It made the bird look like a Bonneville salt flats racer, and for a while after that we used the 'Bonneville Racer' as a code name.

"It was a viable candidate configuration because one of the things we were looking at was beating the radar reflectivity problem by cleverly shaping the airframe rather than by simply using the same methods we had adopted for the 147.

"We had to put the project on the back burner but kept working away at it as we learned more and more about radar shaping. From time to time, we introduced the configuration into our continuing discussions with the Air Force, but were unable to find a specific spot for it in their plans and funding capability.

"**U**NDER THE CODE NAME 'Red Book' we finally put together a formal proposal which used the company designation model 150. We'd had some pretty interesting feelers from the CIA which apparently wanted to operate a completely black program divorced from everything else including the Air Force activity with our 147 recce birds.

"Bill Rutherford and I made the presentation but we were very nervous talking with the agency about the same kind of program we already had going with our established customer, the Air Force. Later we learned that the CIA people with whom we were dealing were also working with the Undersecretary of the Air Force, but that didn't minimize our prob-

lem appreciably so we suggested that if the CIA wasn't responsive within 30 days we'd feel free to deal direct with our normal Air Force channels

"Within a week the CIA responded negatively, suggesting however that we now take the project to the Air Force. We did, after changing the code name to 'Blue Book' which sounded a bit better and less suggestive of a project which might be aimed at Red China, for example. By now it was clear that the project would be under the operational control of the Air Force with the CIA being the user of the intelligence information.

RYAN'S
T. CLAUDE RYAN

Richard Stauss

"Finally we felt the time was right for an all-out effort by unveiling the proposed vehicle through SAC Headquarters in Omaha. The top man there was General John D. Ryan, and we decided we'd match him with our top man, T. Claude Ryan. Between the two Ryans we knew we'd draw a good audience.

"At that time, our project head was William Clasen, a colorful and very knowledgeable former Marine Corps fighter pilot type. We took him along because we wanted a good technician to operate the projector for our slide presentation. Who would be more conscientious about the quality of the presentation than the project engineer?

Teledyne Ryan

THE MARINES'
BILL CLASEN

"The presentation drew the expected SRO audience — standing room only — even for full colonels. We got a very receptive hearing and we knew then and there we had a potential winner. The project was being handled in a very special internal system in the Air Force, but it soon became clear that it was going to have to go competitive.

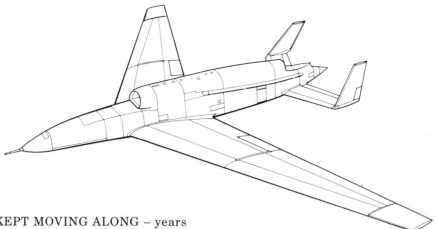

"TIME KEPT MOVING ALONG – years actually – as we moved from the model 150 to the 151 and on through several more proposals to the final 154 configuration.

"Some very SAC-oriented people at North American Aviation got wind of what we were up to and set up a separate division to try and go after the new, higher-flying reconnaissance drone business. North American's guy was John Popp and soon he and I were running into each other at airports and hotels as we continued our contacts with the customer. Later, Northrop tried to get into the action, but the Air Force held it down to competition between Ryan and North American.

"Basically, the mission requirements in range and altitude were pretty much what we had proposed. We'd studied the field and had been operating drones for years so we had the inside track.

"One of the toughest areas was selection of a power plant. General Electric was proposing an engine, but so was Continental Aviation and Engineering, then a subsidiary of Continental Motors which was owned by Ryan Aeronautical Company. To complicate matters, Bill Rutherford, a Ryan vice president who was helping shepherd the project through top governmental channels, was also president of Continental Motors.

"We found we hadn't done all our homework well enough and that North American looked more and more like the winner. During the months that followed there was much activity back and forth and all across the customer spectrum. Even the specifications kept changing. For example, one configuration would go much higher but fly only half the distance. Was this what the customer wanted?

"THE VARIOUS PRESENTATIONS involved decisions by such high-level people as General Steakley, Dr. McMillan, and Dr. Flax of the Air Force, and of course, Bill Rutherford and Ryan President Bob Jackson.

"Right in the middle of the competition we found that the performance characteristics given to us by General Electric on their engine were different than the ones they had given North American. We screamed foul. The meeting which followed was one of the most difficult in which I have ever participated. We ended up having to change the wing configuration at the last minute to obtain a better match with the engine.

"Although the government would be furnishing the engine, we were asked to choose between the GE and Continental power plants — a tough decision from a practical point because Continental was our company. Our position was that we would show what the 154 would do with each engine and the government should make the selection on the basis of the performance of the vehicle. They said 'no dice' and would not allow us to straddle the fence on that one.

"On the basis that General Electric was actually further along in the development cycle we opted for their engine. Rutherford, having helped sell the project, had some pretty strong feelings about the matter as did my other boss, Bob Jackson.

"Even the technical people at Wright Field who would not permit us to straddle the matter of engine selection, ended up doing so themselves, sending the matter up the chain of command for final selection.

"In any case, Ryan, instead of North American, was the eventual winner of the 'Lone Eagle' project and the nod for the engine went to General Electric instead of Continental. It was a pretty wild and furious competition, but at the same time reasonably friendly.

"The second-generation recce drone would have its day and this one would be designed from the ground up rather than being a modification, as was the 147, of an existing drone system."

THE FIRST EAGLE ON WHEELS is moved slowly *to the launch aircraft.*

THE 'COMPASS ARROW' BIRD

A COMPLEX, WELL-PLANNED machine as tailored for the 'Compass Arrow' mission, the 154 pushed the state of drone art just about as far as it could then go.

What did it take to satisfy the Air Force operational requirement to perform a high-altitude, long-range, photographic reconnaissance mission in a hostile air-to-air and surface-to-air threat environment?

To minimize vulnerability in the combat environment, the special purpose aircraft (SPA) operated at an altitude of 78,000 feet and was designed for minimum radar reflectivity and infrared (IR) radiation. In addition, the SPA carried deceptive-type electronic countermeasures (ECM) equipment.

After launch from the wing of a DC-130E director aircraft, the SPA had a powered flight range of 2,000 nautical miles plus another hundred miles during the unpowered glide phase. It flew at Mach .8 during the cruise-climb profile from 72,000 to 78,000 feet. Missions of four-and-a-half hours of powered flight endurance could be flown. The launch aircraft has a range of approximately 2,300 miles when carrying two 154 drones.

Navigational accuracy of one half of one percent of distance traveled can be realized in the primary Doppler/Inertial mode. During flight, the SPA is guided internally by a navigation system which operates in conjunction with programmed data which is loaded into the SPA prior to launch. Excellent flight stability of the 154 as well as navigational accuracy contribute to successful high-altitude photographic missions.

The KA-80A camera system provides 1,720 miles of along-track coverage at operational altitudes with 43 miles lateral coverage and ground resolution of one foot directly under the flight path.

Enemy radars and IR detection devices find the 154 almost impossible to locate and track due to the unusually effective, low reflectivity and radiation design features. Radar cross section of the 154 airframe has been minimized by contouring the structural shapes, shadowing the engine intake and exhaust ducts and by using radiation-transparent and radiation-absorption materials.

Infrared suppression has been achieved by positioning the engine on the top area of the SPA, extending the fuselage aft of the engine, shadowing the tailpipe by the twin-canted vertical fins and using engine inlet air to cool the engine ejector nozzle.

The DC-130E launch aircraft is capable of carrying two model 154 reconnaissance drones or a combination of the 154 and 147 drone systems. After launch at altitudes between 15,000 and 25,000 feet, the computer-controlled vehicle climbs to a minimum of 72,000 feet approximately 300 miles from launch point before starting the photographic portion of the operational mission.

Prior to launch, final checkouts are performed, the SPA fuel tanks are topped off to a maximum capacity and updated mission-planning data are fed into the on-board navigation system. During the initial period after launch, command and control are exercised by the DC-130E.

The 154 is designed to use the MARS mid-air retrieval system in which a Sikorsky twin-turbine CH-3E helicopter snatches the engagement and 100-foot recovery parachutes which are deployed on command or, in case of emergency situations, automatically. The 3,800-pound 154 drone can be recovered by helicopter at altitudes of up to 10,000 feet. The chopper can tow the SPA at 70 knots for 50 miles in returning the vehicle and collected intelligence data to the recovery area.

The drone vehicle may also be recovered by parachute descent to the ground or to the water. In this operation, impact bags are deployed to cushion impact with the surface.

External command and control functions are provided by a microwave command guidance system (MCGS) consisting of equipment in the drone, an air director station in the DC-130E launch plane and a ground director station. The MCGS is used to transmit command and control functions during the recovery phase of the mission and is effective over a line-of-sight distance of 200 miles. The ground director equipment is also able to locate and plot the position of the drone and to receive and display telemetry data from the aircraft.

Unlike the 147 family of drones, the 154 was designed to have a destruct system to prevent disclosure of sensitive on-board equipment.

THE SWEPT, LOW WING of the 154 has a span of 48 feet. The length of the fuselage is 34 feet. The drone is powered by a single General Electric YJ97-GE-3 jet engine, producing 4,000 pounds of static thrust at sea level. At launch, fully fueled, the drone weighs 5,400 pounds. Its unfueled weight is 3,800 pounds.

The fuselage has inclined, flat sides which reduce radar reflectivity to an absolute minimum. For the same reason, the twin vertical tails, which also provide some infrared shielding of the engine exhaust, are also sloped.

To facilitate maintenance and handling, as well as to simplify production, a modular design concept has been used. The SPA separates into five modules – nose cone, fuselage, empennage, tail cones, and wing assembly. The fuselage assembly separates further into forward, center, and aft sections. Similarly, the wing separates into a center section and two outboard wing panels.

Extensive use is made of plastics including radar-absorbent material in strategic areas, not only as a cover over other materials but as a basic primary structure. All fuel is carried in the wing center section, which uses a heat-activated, epoxy-sealing system especially developed for the 154.

Because of the urgency of the Compass Arrow mission, the Model 154 followed a concurrency approach in which the research and development flight test program was conducted simultaneously with the production of operational vehicles.

WHILE THE 'SKUNK WORKS' shop at Ryan got started on fabricating parts for the 154, program chiefs began to tackle the job of managing the project. They had been used to the more loosely administered Big Safari approach on the 147 than to the standard procurement methods of the Air Force Systems Command which now had contract responsibility.

Before long, the program (to no one's great surprise) was running behind schedule. No one had anticipated all the design changes and complexities of building a from-the-ground-up bird, and both Ryan and the Air Force, in time, agreed that initially they had been too optimistic on their timetable.

Schwanhausser recalls, "The 154 was a victim of too much optimism in the heat of a very tough competition to get the business. On our other recce programs, if you look at the track record, you'll find it's been very good.

"We ran into a longer development cycle on the guidance system than we had intended. That was the most difficult one we had to work with.

"Then too, the 154 was probably over-managed. The Pentagon, at the Assistant Secretary level, was involved. Headquarters AFSC also got into the management cycle as did the SPO office at Dayton. Finally, the Air Force put a lot of management into the program right in San Diego at the Ryan plant. The SAC people also wanted to be sure that this

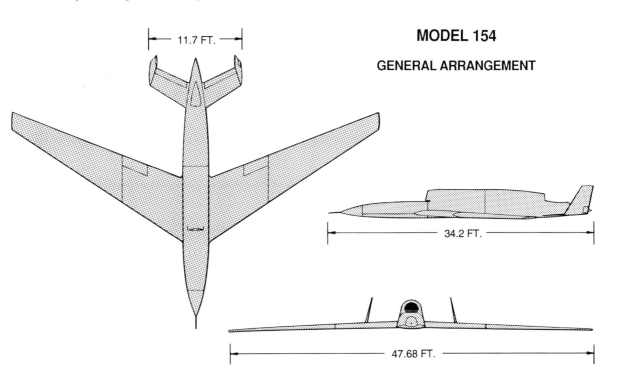

MODEL 154

GENERAL ARRANGEMENT

11.7 FT.

34.2 FT.

47.68 FT.

happened, or that happened, and I would say before we were through it turned into a pretty unwieldy management apparatus compared to Big Safari philosophy.

"We had a monthly program review out here in San Diego, in some cases involving as many as 200 people, and you can't have that many drivers. Fortunately, the CIA didn't get involved in the hardware side of the business; all they wanted was the end result of intelligence flights.

"The nav system was new and in its early configuration required development of a brand new digital Doppler. Originally it was just going to have a heading reference, but at the last minute in the contract definition phase, we switched to an inertial platform and computer. So it took time to get the good reliability that would make it work consistently and accurately. In final form, it worked well.

"Many of the flights during operational testing and evaluation far exceeded specifications on accuracy, and on the average, we have been well within spec. While the bad contract numbers were certainly there in terms of taking longer and costing more, we had one of the best guidance systems in any airframe flying at that time.

"Later there was some bad image, too, from the fact that Teledyne Systems was providing the inertial platform and computer, but they were selected at a time long before there was any connection between our company and Teledyne, Inc., which didn't buy Ryan until 1969. That developed after the fact.

"In the long run, people remember performance; not the woes you had along the way. But everyone's in deep trouble if you have the development and financial woes and then don't perform. Fortunately, the 154 was a fine bird and could have performed even better in the future."

TESTING, TESTING, TESTING

IN ALL, SOME 28 MODEL 154 aircraft were built with the first production vehicles coming out of the shop early in 1969. There was one STV (static test vehicle), two CTVs (captive test vehicles), five FTVs (flight test vehicles) and 20 production vehicles, designated P-1 through P-20.

The flight test program got under way at Holloman Air Force Base on June 4, 1968 when the C-130 launch plane carried two birds, CTV-1 and CTV-2, aloft for the first 'captive' tests of the drones without engines installed.

Not all of the CTV flights were captive. In many tests the drone was released for free,

FIRST OF TWO CAPTIVE TEST VEHICLES is ready *to leave the factory production line for transport to Holloman Air Development Center.*

unpowered flight in order to check launch characteristics, separation from the C-130 and parachute recovery system operation including both mid-air retrieval with the MARS helicopter, and ground landing using the impact bags. To properly simulate the bird at launch an 'iron horse' structure was used as ballast to represent the engine weight.

At the same time, tests of the recovery system were being conducted at the El Centro parachute station where towing and docking by the MARS retrieval helicopter was being checked. Later at Pt. Mugu the hulk of the 154 static test vehicle was used for MARS training.

Teledyne Ryan Aeronautical

Without the navigation system installed, the first powered free flight was made at Holloman Air Development Center on September 10, 1968. In the series of FTV flights which followed, an evaluation was made of aerodynamic characteristics, basic performance, stability, and powered launch procedures. These flights were restricted to the confines of the White Sands Missile Range until confidence had been built up in the system.

During the following year, 1969, a total of 152 captive flights and 42 free flights were made, using seven instrumented vehicles. It was the start of the long, continuing testing period necessary for development of such a complex reconnaissance system.

In April '69 FTV-3 (flight test vehicle number 3) was expended when the tow line of the mid-air retrieval helicopter was severed as the 154 was being reeled in.

Two more vehicles were lost to the test program in May. FTV-5 experienced erratic control maneuvers immediately after launch with resultant descent and impact. The damage was so extensive as to be not economically repairable for further flights.

FTV-4 was expended during recovery. The vehicle risers were severed after being engaged by the MARS helicopter and resulted in destruction of the vehicle upon ground impact.

In July, P-3 (production vehicle number 3) was lost on its first free flight due to control problems.

Then came August and the attention-getting parachute descent of P-4 when it made an unscheduled off-range recovery at Los Alamos due to a control actuator failure.

FLIGHT ACTIVITIES thereafter, were restricted to within the WSMR range boundaries as efforts were concentrated on establishing a reasonable confidence level of SPA reliability to satisfy WSMR range safety requirements.

Meantime, use of the MARS recovery system had been restricted in favor of the impact bag configuration for ground recovery until the mid-air retrieval technique could be investigated and improved.

Following a successful on-range navigation mission by P-5 in September, the range restriction in effect since August 4 was lifted.

A PAIR OF BIG-WING COMPASS ARROW BIRDS (top of page), *fit snuggly on the pylons beneath the wing of a DC-130 launch plane.* **Below:** *Ready for take-off.*

ledyne Ryan Aeronautical

Subsequently, two good long range nav flights were accomplished with P-4, the SPA which had made headlines at Los Alamos.

Its November 21 flight on an off-range navigational mission nearly caused the 154 program to make the papers again, as related by Dale Weaver who was aboard the C-130 director aircraft:

"Due to a circuit failure, the bird went into the automatic recovery mode. We were over northern Arizona. It was winter and there was snow on the ground as the 154 came floating down onto a remote section of the Navajo Indian reservation.

"By the time we were circling overhead in the C-130 we could see a fair number of people around the drone. We buzzed the area to warn the pick-up truck, horses and people away. The 154 looked to be in pretty good shape even though it had impacted into some small trees.

"The closest field where we could land was Farmington where I made arrangement to get some rental cars but because of the crowd gathering at the scene we thought we had better get somebody there immediately. We called Durango, Colorado, and had a rental helicopter come to Farmington to pick up Captain Jackson and me. The others followed by car.

"Based on the Los Alamos experience we put into effect the contingency plan which called for a group in CH-3 helicopters from Holloman to fly out to where the drone had landed.

IMPROMPTU RESCUE of a dismantled bird *is made in the cold dawn on a Navajo Indian reservation.*

Teledyne Ryan Aeronautical

"When Captain Jackson and I arrived in the rental helicopter we were pleasantly surprised to find that the Navajo reservation Indian police had secured the area and had everything well under control.

"In the group was an ex-Air Force sergeant who had recognized the bird for what it was 'a special purpose' machine. It was he who had first contacted the reservation police, saw that a perimeter guard was established, secured the recovery parachute and otherwise took charge in a very professional manner.

"We choppered out to the nearest trading post where I was able to get on the phone to San Diego and to Holloman to let them know what we were doing. Then it was back to the downed 154 where we spent a pretty long, cold night standing guard. Fortunately the reservation police kept us pretty well supplied with hot coffee.

"Shortly before dawn the wrecker and flatbed truck arrived. We took off only the wing tips, cleared a path and then walked the bird out to where we could get it on the flatbed. Reservation police escorted us all the way to the edge of the reservation at Farmington where New Mexico State Police met us to escort the extra wide load on down to Holloman at Alamagordo.

"I hope the written commendation which went to the Navajos and the ex-Air Force sergeant reached them in due time through official channels. They did a really professional job; and certainly the unscheduled landing of the 154 was a unique event for them."

THE 154 WAS IN THE MIDST of an 18 month de-bugging period. Beset with problems of control surface actuators, the MARS system, premature engine run-downs, fuel tank leakage and other chronic malfunctions, Teledyne Ryan and Air Force engineers managed to identify and solve them one by one.

Additional tests had to be made of the MARS system including the parachutes deployed by the 154 as well as techniques of reel-in, tow and docking after the CH-3E helicopter had snatched the bird in mid-air. The equipment, which had been restricted after the May 29 loss of FTV-4, was finally reinstated after an additional test program at Pt. Mugu with two "hulks" (the static test vehicle and FTV-5, refurbished after its crash in May).

PILOTS OF SIKORSKY CH-3E twin-turbine choppers *were carefully trained to recover the 3,800 pound unmanned vehicle.*

But one problem refused to go away. The extremely sophisticated navigation system so necessary for the success of accurate photo reconnaissance was being stubborn. Larry Emison describes it:

"The primary Doppler/Inertial navigation system called for an accuracy of 1/2 of one percent of the distance travelled, equivalent of an error of no more than 5 miles for every thousand miles flown. To achieve such precise navigational accuracy, usually after many course changes en route, was no small task.

"To assure reliability, the system had four back-up modes of operation, but to provide this capability added to the technical complexity. Further, radiation from the on-board elec-

RADAR TARGET SCATTER (RATSCAT) Facility *at Holloman Air Force Base provided means to analyze radar return from high-flying Model 154 RPV.*

tronic equipment had to be kept at an absolute minimum to avoid detectability by enemy radar.

"The heart of the whole system is a miniature computer looking at Doppler velocity, looking at inertial velocity and constantly comparing them to update navigational accuracy. The Doppler and the inertial platform send 'state of health' information to the computer and to the built-in test equipment (BITE). They continually monitor themselves and if they have any problem they issue a signal to the computer, and the computer takes them off the line.

"Let's say you are flying on the primary Doppler/Inertial system and you lose the inertial platform, the heading and attitude reference system which we call HARS.

"When that happens, the computer switches over to take heading information from a directional gyro to compare with the Doppler velocity signal. This is called the Doppler/Directional Gyro mode. It degrades the accuracy we have with Doppler/Inertial but it is still a good, automatic back-up system.

"Similarly, if we lose the Doppler, we rely on the inertial system for heading and velocity information. Again there would be a slight down-grading of navigational accuracy.

"Again, if we lose both Doppler and the inertial platform we get heading information from the directional gyro. The computer then provides navigational equations based on the last credible velocity information that it received from either the platform or the Doppler.

"Then, should we lose the computer, which does the calculating and monitoring of the other systems, the vehicle will turn around and head toward a predesignated recovery spot on the directional gyro.

"So, with five separate modes of operation we assure ourselves that the bird can get home. It's not difficult to understand why qualifying this system was a very complex deal.

"Before the DC-130E launch plane ever leaves the ground, the memory of the 154 on-board computer is loaded with equations on the route to be flown in terms of specific points of latitude and longitude where course changes will be made.

"As many as 38 lat/lon changes in the flight route can be loaded into the tape cassette which the launch control officer takes aboard with him. Before launch he loads that particular route program into the 154 computer from his station aboard the launch plane out to the SPA on the wing pylon. Even at the last minute before launch the commander can have an entirely new flight plan loaded into the computer.

43

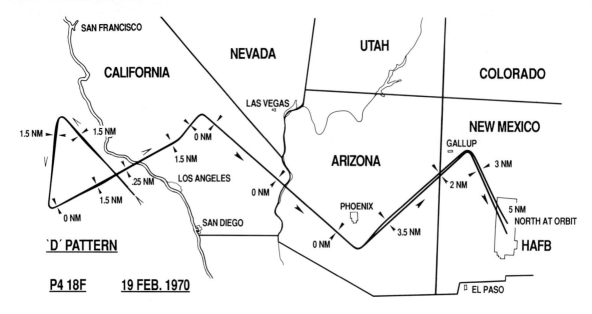

San Francisco

CALIFORNIA
NEVADA
UTAH
COLORADO

LAS VEGAS

NEW MEXICO

1.5 NM   1.5 NM   0 NM
GALLUP
3 NM

1.5 NM
ARIZONA
2 NM

V

.25 NM   LOS ANGELES

1.5 NM
0 NM
5 NM
NORTH AT ORBIT

0 NM
PHOENIX

SAN DIEGO
3.5 NM

`D´ PATTERN
0 NM
HAFB

P4 18F   19 FEB. 1970
EL PASO

NAVIGATIONAL ACCURACY with errors of less than 0.3 percent *was demon-strated on flight air-launched over the Pacific and flown in a zig-zag pattern across California, Arizona and New Mexico.*

"Once launched on its course, the bird's computer will constantly compare the actual track with the programmed track and keep the vehicle on course. The cassette tape also programs at what lat/lon coordinates the MCGS airborne receiver is turned 'off' and again 'on' before recovery. Control remains with the automatic navigation equipment until the program tells the computer, 'now turn the MCGS back on so ground control or airborne control in the C-130 can take over remote control and bring the bird home manually.'"

CHECKING OUT and confirming the accuracy of the nav system was a time consuming and often frustrating process. Using FTV-4 and FTV-5, a series of navigation system checks was run on three off-range patterns.

Navigation pattern A went beyond the White Sands Missile Range but still within the relatively remote northwest corner of New Mexico. The C pattern took the 154 into Utah past the Four Corners area where the states of Utah, Colorado, New Mexico and Arizona meet.

The D pattern represented a real graduation exercise and was the one which aroused the interest of Congressman Udall of Arizona.

Three navigational proving flights were flown over the D pattern early in 1970, nearly a year after the controversy had arisen about the overflights from California to New Mexico.

Taking off from Holloman Air Force Base, the launch C-130 aircraft carrying two 154 SPA was flown to the Pacific Missile Range off Pt. Mugu. There the 154 was air launched on a northwesterly course paralleling the coast, followed by a left turn which took the bird a hundred miles out over the Pacific before the course was changed for Edwards Air Force Base. From there the route was southeasterly to the Army's Ft. Huachuca, then northeasterly and finally back on to the White Sands Missile Range.

During the flight the recce drone was monitored by the Federal Aviation Administration so there would be no interference with or danger to other air traffic, whether commercial, private or military. After launching the 154, the C-130 'escorted' the 154 back to Holloman on a more direct course as an in-flight command post to take control of the bird at any time should that be necessary.

At Holloman, the entire operation was under surveillance at all times through a radar-link network which constantly updated and displayed the progress of the 154 flight on a plotting board. Additionally, a back-up test group with complete instrumentation and control capability was stationed at Luke Air Force Base, near Phoenix, at a point mid-way on the overflight from Pt. Mugu to Holloman. Aircraft and helicopters were also stationed at Luke in case the bird went down in some remote area so as to get to it in a hurry for security reasons.

Teledyne Ryan Aeronautical

RESPONSIBILITY FOR THE SUCCESS of unmanned missions rests in the capable hands of the direct control operator at the console aboard the launch plane. Ready to send the 'bird' in photo at right on its way is DCO Walt Hamilton of Ryan.

Navigational performance of the 154 on all three proving flights on the D pattern range was excellent. One plot showed a maximum off-course deviation of 10 miles near the end of an 1800-mile track.

"On one of the flights on the long D range," Larry Emison recalls, "we lost the inertial platform so that it automatically went to the Doppler/Directional Gyro back-up mode. It did a beautiful job of maintaining navigational accuracy."

GIVE 'EM THE AXE, THE AXE. . . .

BEFORE THE AQM-91 could be placed on a 'ready to deploy' status, an operational test and evaluation of its suitability to fly SAC reconnaissance missions in an operational environment would have to be made.

The program for the 100th Strategic Reconnaissance Wing to conduct the tests was assigned the nickname 'Busy Robot' and was accomplished between August and December, 1971, only months after President Nixon's overtures to China.

The eight-mission test schedule was staged out of Davis-Monthan Air Force Base with launch and recovery accomplished at Edwards Flight Test Center. Range coordination and MARS helicopter support were provided by the 6514th Test Squadron. After recovery the drones were uploaded at Edwards and returned to Davis-Monthan for regeneration. The eight test missions were flown over California and Arizona on navigational routes which involved flights of 3-1/2 to 4 hours during which from 1600 to 1800 nautical miles were covered at altitudes up to 81,000 feet. At conclusion of the tests, the final report stated:

"Operationally, the AQM-91A is capable of performing high altitude photo reconnaissance missions of 4 hours flight duration, 3.1 hours
camera duration, and 1900 nautical miles in length. The Guidance Navigation System (GNS) is capable of executing all programmed functions and directing the Special Purpose Aircraft (SPA) over the planned track with a high degree of accuracy. In addition, because of outstanding vehicle stability, higher photo resolution and improved photo overlap consistency is achieved. Together the AQM-91A and KA-80A scorer provide an excellent unmanned strategic reconnaissance system."

"GNS performance was excellent and in the primary Doppler/Inertial mode navigational accuracy of less than one-half of one percent of distance traveled can be expected."

Two flights recorded navigational errors of only 0.3%.

While the OT&E confirmed the readiness of the AQM-91A Compass Arrow vehicle and its systems, the report noted those support areas which needed improvement. Mainly these involved availability of MARS recovery helicopters with greater performance and the need for uniform configuration of DC-130E launch/director aircraft.

Evaluation of the suitability of electronics gear was a major consideration, and test flights no. 6 on October 20 and no. 7 on November 4 were conducted in connection with Electronic Warfare Tactical Environment Simulation (EWTES) at the Naval Weapons Center, China Lake, California.

The Navy installation included a variety of simulated Soviet SAM systems including Fansong B and E radars and associated equipment of SA-2 Guideline missiles.

The objective was to test the AQM's capability, with its low radar cross-section, to reduce the probability of detection by the enemy, as well as tests of the jamming system which was designed to counter a variety of SAM systems if the vehicle was detected.

The specific objectives were three-fold: To measure the effective radar cross section of the AQM-91A drone; to determine capability of countermeasures equipment to provide vehicle self-protection in a hostile missile environment; and to test the capability of inflight receiving equipment to record enemy threat conditions.

Tests of the special purpose countermeasures package used as a penetration aid against SAM systems' radars were not fully completed due to malfunctioning equipment.

But one thing was eminently clear. Because of the drone's small effective radar cross-section it was obvious that enemy SAM radars would have a difficult time finding it, and therefore its vulnerability in a hostile atmosphere was minimal. In fact, EWTES radar operators found it almost impossible to acquire the drone with either the Fansong B or E without very accurate space-position information on the vehicle being provided.

The AQM-91A had passed its tests, and was ready to deploy. But with the war in Vietnam winding down, where would it find a mission?

EVEN AFTER COMPLETION of the successful operational test and evaluation, the 154 seemed doomed to a state of suspended animation. The capability was there, but it no longer had the strong mission for which it was originally planned.

Long before President Nixon's dramatic announcement in July 1971 that he would visit Peking, steps were quietly being taken behind the scenes toward a rapprochement with Red China.

The landing of a strange new pilotless jet plane anywhere in mainland China or in territory of a country friendly to Mao would hardly be conducive to the more relaxed climate the administration was seeking.

"Our bird," explained Schwanhausser, "was really configured for China, and now we're playing ping pong with them! It was designed to go into an area where you must have very precise navigation and very high altitude so the vulnerability would be very low to interception by manned aircraft or to being knocked down by SAM missiles. Before long there was no great desire to get the capability fielded; and it would have been a shame to waste it in North Vietnam where the 147SC was doing the low-level mission with great success.

"The 154 is ideal for the side-looking radar mission and some other vehicles like that were appropriate and could indeed have been used in the Middle East."

Timing in the introduction of a new system is all-important; but the 154 unfortunately missed its place in time. "Had it been completed two years earlier," Bob Reichardt states, "I think it would have long since been operational. The changes in the political scene, particularly in China, made all the difference and by the time we were ready it was not politically feasible to operate it.

"In December 1971 the 154 completed its OT&E phase and was ready for deployment. Within the military intelligence community SAC announced its satisfaction with the system and that they were ready to go operational with it, but the political restraints keep it on only on a stand-by basis ready for use when an acceptable mission was established.

"As far as the hardware, people and operating command were concerned, the system was ready to go.

"In the spring of 1972 there was a possibility of a mission over Cuba but that did not come to pass.

"If the 154 had gone operational it would have been a significant milestone – a tremendous step forward."

"The only real hurdle holding up the 154 program," says Schwanhausser "was the decision to deploy the system some place and put it into operation. Nothing is proven by test and evaluation programs; the capability is only demonstrated in day-to-day operational use in the field. The 147 bird had a theoretical capability for two years; it was only with the Gulf of Tonkin incident that it was given a chance. We'll not know about the 154 and similar birds until they too are put to work.

"In this respect, recce drones follow the pattern of the helicopter. We all assumed it would some day be a military workhorse but it was not until Korea that the 'helo' was given a chance; then the military found more chores for it than had ever been envisioned.

ON STANDBY, READY TO DEPLOY. *But on what mission?*

"Another capability which demonstrates this fact is that of V/STOL aircraft. Many research planes, many technical approaches, have demonstrated unique capabilities.. But until someone buys a quantity of advanced design aircraft and puts them to work we will never know their real potential.

"Our past experience has been that we don't really get going on a program until we have demonstrated operationally what we can do and someone has come back with the bacon. You bring the pictures back from the field and take them to the military chiefs; they like the results they see and they know you have done what you said you could do. Then the odds of continuing the program and building some more birds to satisfy a very real need with a validated system are pretty fair.

"Perhaps the 154 and its successors will be like the 147 in that these birds almost have to wait for some international event to trigger their operational use.

"Any production response right now [1970] to an additional 147 requirement would be very rapid, but any response to the 154 would have to be much longer. The engine, for example, is not in production, so that a reorder would mean the expense and delays of again starting up the engine line. And the price would necessarily be high. We at Teledyne Ryan can build and bend the airframe hardware a heck of a lot faster than GE can build engines. And the guidance system would again be somewhat of a pacing item.

"Unfortunately rather than having an assignment, the 154 was placed on a ready-alert basis at Davis-Monthan waiting for that assignment."

THE 154 WAS DESIGNED with unmanned overflights of Red China in mind, but with the rapprochement initiative by Kissinger-Nixon in 1971 it was considered prudent not to use Compass Arrow for that mission. Also, the 154 bird was suitable for missions in connection with the Yom Kippur War in the Middle East but, again, the political aspects prevented serious consideration of that option.

What then did happen to the 154s?

A number of stories have surfaced, the most credible being that the highly classified vehicles existence 'and possible missions' should be kept secret to safeguard their technical innovations. The best way to maintain that secrecy probably was to guillotine the airframes at the Air Force facility at Tucson, Arizona. That is the story we most frequently hear together with the report that only the engines were salvaged.

As one knowledgable individual at Dayton phrased it, "The 154 was an embarrasment to the military-industrial establishment because by the time the money and effort had produced a vehicle of good potential there was no requirement for it, and that was an embarrasment, so the birds were melted down except for the engine."

ISRAEL'S 124I

WHILE ONE TEAM of Ryan technical representatives was busy across the Pacific in the Vietnam theater, another team was heading east across the Atlantic and Mediterranean to Israel, introducing the first unmanned reconnaissance 'birds' to the Middle East.

TEL AVIV, FALL, 1970 – An uneasy peace, negotiated by U.S. Secretary of State William Rogers, which became effective August 8 as a follow-up to the Six-Day War of June 1967, continued between Israel and Egypt along the Suez Canal. New efforts were under way to try and get both sides to move back from the canal as a means of further reducing tension.

How would such a peace be maintained; who would see to it that neither side took advantage of the situation?

Reportedly a U-2, possibly flying from Turkey two days after the cease-fire, found that Soviet missiles had been brought into the 31-mile truce zone along the canal in violation of the agreement.

Secrets have a way of not remaining secrets very long and it was thus that readers of the Jerusalem Post on November 1, 1970, learned from a dispatch originating in the Daily Express of London that "Israel is interested in an American robot plane, the Firebee, to counter the missiles on the Suez Canal."

The Daily Express continued –

"A correspondent, Ivor Davies, reported from Florida that he watched the Firebee being put through its paces at Tyndall Air Force Base. He says the plane can be programmed to do almost anything –from knocking out missiles to acting as spy plane.

AT AN ARMY LAUNCH PAD IN NEW MEXICO, Israeli observers *evaluate the potential recce value of the Firebee.*

Ryan Aeronautical Company

"Recently a team of Israel Air Force officers and aerospace experts secretly spent five days at the California plant of the Teledyne Ryan Aeronautical aerospace firm which makes the aerial wizard.

"With unerring computerized accuracy defense teams [at Tyndall] have destroyed other Firebees [target planes] masquerading as attacking bombers."

The Daily Telegraph, quoting Maier Asher in Tel Aviv, reported three days later that the Firebee "is considered to be one of the best currently available answers to the SAM-2 and SAM-3 missiles of the kind set up by the Russians in Egypt." The London newspaper also observed that hostilities between Israel and Egypt might be resumed 'on Thursday' on expiration of the cease fire.

THE ISRAELI GOVERNMENT'S interest in reconnaissance drones managed to stay pretty well underground and out of the news until July 6, 1971, when Jack Gould wrote a piece for the New York Times headlined

U.S. IS TESTING TV ON DRONES AS WAY TO PATROL SUEZ CANAL

The article brought back memories of President Dwight Eisenhower's 'open skies' proposal at the Geneva Conference of 1955. With the U-2 soon to take the skies, the President then had offered a plan of mutual aerial surveillance as a means of detecting any potential surprise attack by either East or West.

Nikita Khrushchev was not interested. Nor was he interested five years later when, after the downing of Francis Gary Powers' U-2 over Russia, Eisenhower again volunteered to donate U.S. reconnaissance aircraft for operations under United Nations sponsorship.

Reported the New York Times [whatever the accuracy of its source] in relation to the similar problem in 1971 along the Suez Canal:

"Tests of live TV as a means of maintaining continuous airborne surveillance of the Suez Canal are being conducted by the U.S. Air Force along a strip of the California coast. Unmanned aircraft are launched by planes and then controlled by computers to fly back and forth over fixed paths and transmit pictures of West Coast terrain to ground observers.

"Word of the project was learned over the weekend at the United Nations, where there is substantial sentiment that the new peace-keeping instrument be made available to the world organization.

"If applied to the Middle East, the system would relay views the length of the Suez Canal to a receiving station near Tel Aviv. By tape-recording the pictures, Israeli military officers could make instant comparisons of activity along the Canal and determine whether the similarities warranted response. The live pictures could also simultaneously be relayed by Air Force satellite to Washington, where the Defense Department could immediately examine the visual basis for any incident.

an Old wine ...

L. Col. Manoff & Team.
July 1785

David Packard, United States Deputy Secretary of Defense, has witnessed the California demonstrations, which would technically enable Israel to withdraw her forces from the Suez banks while retaining a constant electronic watch over both sides of the waterway.

"State Department representatives are in Cairo trying to arrange an interim reopening of the Suez. The Soviet Union and Egypt, however, insist that Israel first withdraw her forces from all occupied Arab territory."

SEEKING AN ANSWER

ISRAELI INTEREST in the intelligence-gathering potential of aerial training targets which they could modify to carry cameras on photo missions over 'enemy' territory had its genesis in the mid-60s. After all, sporadic fighting had continued along Israel's Egyptian, Jordanian and Syrian borders for almost a decade.

Interested in also improving their ground-to-air missile capability against potential enemy aircraft, the Israeli Air Force sent a group of officers to the United States in March 1967, three months before the Six-Day War, to learn more about Ryan Firebee targets for use in weapons system training.

That fall, Ryan sent a technical group to Tel Aviv to further discuss Firebee target systems, but on this occasion the Israeli officers expanded the discussions and hinted at a requirement for reconnaissance.

"The question of unmanned aerial reconnaissance," writes Yoash Tsiddon-Chatto, then Chief of Planning and Operational Requirements of the Israeli Air Force, "had been raised with a certain degree of urgency in 1965.

"The introduction of SAM-2 missiles in Egypt and Syria created a serious threat to our intelligence-gathering capability. Having at that time no access to U.S. photo-reconnaissance technology, we examined the possibility of using the twin-engine French Mirage IV. This manned aircraft had a sustained Mach 2 capability of 20 minutes which should have allowed it to penetrate the intersection of two SAM-2 defense circles. But operational research proved our assumption to be futile. A high-altitude U-2 type of operation appeared to be a better solution.

"I later approached the Chief of the Air Force, Motty Hod, and proposed evaluating, and maybe acquiring, license for the Nord Aviation R-20 drone.

"Motty was absolutely against the introduction of drones in the I.A.F. and so was Oded Erez, then Deputy Chief of Staff. They considered drones an affront to a fighter pilot's prestige.

"After presenting my case to the Chief of Staff, Lt. Gen. Haim Bar-Lev and to Maj. Gen. Eliahu Zeira, then Chief of Intelligence, they became very keen on drones.

"The result was a letter of general interest I received from the Purchasing Department of the Ministry of Defense the end of 1969, signed by Yacob Shapiro, stating that the Ministry would be ready to evaluate a proposal.

"I went to France to work out an agreement between Nord Aviation and Cyclone Aviation, a new Israeli firm of which I had become managing director, to obtain French R-20 drones.

"A formal proposal was submitted to the Air Force on behalf of Cyclone, but Motty Hod was apparently fed up with the French and suggested that if he's going to have drones we should try to reverse the U.S. export license limitations since the Ryan Firebee type vehicles provided far superior performance. 'Shapik' Shapiro, as head of Defense Procurement, was also very pro-U.S., pro-Ryan since his days as purchasing head for Israel in New York."

THE POSSIBILITY OF THE ISRAELIS purchasing Ryan Firebees lay dormant for over two years. Some contacts with the Israelis were kept open by Bob Schwanhausser, including a visit in August 1969 when he was accompanied by R.A. (Pete) Petrofsky who went along "for the technical side and my experience with the 147 recce programs on which I had worked."

Then, in March 1970, Schwanhausser, accompanied by Bernie Paul, was in Southern Europe on a business trip, following up on the company's long-time effort to obtain a contract to operate the NAMFI Range – NATO Missile Firing Installation – on the island of Crete.

Why not drop over to Israel? As Swany said, "We wanted to go down just to see the place and meet the people."

The timing was excellent.

As the Israeli Air Force later wrote, "Towards the end of the Suez attrition war in the middle of 1970, several [manned] Israeli Phantom aircrafts [flying photo reconnaissance missions] were shot down by Egyptian missile batteries. Consequently, the IAF searched for an aerial vehicle which was capable of running intelligence operations and not jeopardizing human life. The solution was the RPV."

"**W**E ALREADY HAD some high level contacts with military procurement people in Tel Aviv," Swany explained. "This had resulted from business dealings of the company's subsidiary, Continental Motors Corporation. Meanwhile, G. Williams (Bill) Rutherford, who was in on the 147 reconnaissance program in the early days, had moved up to President of Continental.

"He'd been selling Continental engines to Israel for their military tanks as they sought to maintain the balance of armaments in the Middle East. Business arrangements on the tanks had been handled through 'Shapik' Shapiro, then head of the Israeli Purchasing Commission in New York.

"I'd met Shapik at the time and briefed him on target drones, but since then he'd returned to Tel Aviv in charge of all Israeli defense procurement. Shapik and his successor in New York, Samuel 'Bondi' Dror had shown a lot of interest in drones but security regulations prevented us from talking about the bird's reconnaissance capabilities. They had a light in their eyes, but no money in their budget.

"Still they invited us down since we were in the eastern Mediterranean anyway in hot pursuit of the contract to operate Ryan target drones on the NAMFI Range.

"We didn't have anything to present formally, other than straight target drones as represented by our 124 model, the same as the training target known to the U.S. Army as the MQM-34D. Of course the 147 series of recce birds was in our minds, but we carried no technical information on this capability.

VISITORS TO ISRAEL

"**T**HE RECEPTION from the Israelis was something I'll never forget. For years I've banged around the world in the role of 'salesman' trying to score points with potential customers by playing the gracious host. The Israelis completely reversed the roles.

"We were met by a chauffeur-driven car and a major who had been assigned by Shapik to look after us throughout our stay. Our potential customer was hosting us!

"Next morning we were taken straight to Shapik's office. He made it clear they were seriously interested this time, and that we were to work with Lieutenant Colonel Uri Talmor who was waiting to receive us.

"Here we got another surprise, for Uri appeared in his office in flight togs and jacket having just come in from a flight in the rear

seat of an F-4 Phantom jet. We learned that staff officers held two jobs – one in management in their version of the 'Pentagon'; the other an operational assignment in a squadron. They'd fly a regular mission in the morning, then report in the afternoon to their desk assignment. It is little wonder that Uri understood the practical aspects of his job in Military Requirements.

"When we arrived Uri excused himself to take a phone call; the conversation was in Hebrew. As he hung up he turned around with a big smile and apologized for the interruption. 'I am sorry, gentlemen, to have delayed you but we just got two MiG's and that is very good news!'

"If and when any drone business materialized we had expected to work with Israeli Aircraft Industries, but we learned differently when Uri asked a civilian to join our discussion. He was Yoash Tsiddon-Chatto, by then managing director of Cyclone Aviation Products, Ltd., a new company which was just being set up.

"Israeli Aircraft had too much work already on hand, we were told, and that Shapik suggested we work with Cyclone. So, when the chief of procurement says, 'please work with these people,' you do so. Yoash knew we were coming, and through other contacts we knew he might be involved.

"Our talks went on for nearly two hours and it was a tricky course we had to run trying to satisfy their 'how-do-you-know-you-can' inquiries about reconnaissance while fighting the security problem.

"We were able to do this by extrapolating the information we had on the 124 target system. Using a blackboard I described what I thought was in the realm of feasibility for flying high and flying low using the 124 as a basic vehicle. Without mentioning the 147 I told them how we could get improved performance by extending the fuselage, using a larger wing, putting on a new nose for photo missions, and flying this or that piece of camera equipment.

"Like everyone else who had read the newspapers, their intelligence people undoubtedly had briefed them on the 147 missions in Southeast Asia, but they didn't appear to have any really concrete information. However, as one of the officers flipped through a file I saw that he had some pictures taken by a Hycon 338 camera which must have been shot by one of our 147 birds .

"Next day we got down to details and costs. Working half way around the world, time schedules are all shot to hell, so while we slept in Tel Aviv, our technical and cost people worked on the problem in San Diego.

"Meanwhile we helped draft the formal proposal, and three days later – with all the information finally in from San Diego – it was submitted by Cyclone to the Israeli government. Then Bernie Paul and I packed up to go back to Greece to resume our efforts to win the NAMFI contract as that was our prime reason for being in the area."

IT'S A DEAL

"**W**HEN WE GOT TO ROME we had a phone call waiting there asking us to return to Tel Aviv for more conferences. This resulted in several proposal options being submitted by Cyclone. Between conferences, our aide, the major, escorted us as guests of Israel on a tour of the country including the Golan Heights, Sea of Galilee and Jerusalem. On our departure the Israelis accepted our invitation to visit the United States and continue technical discussions there.

"When they arrived in San Diego in June 1970, they brought a team headed by Uri Talmor as chief of requirements. On the team was Levi Tsoor, an F-4 Phantom wing commander, under whom Uri flew on squadron missions. But on the requirements assignment over there, Uri was his boss. The Israeli group visited Firebee target operations at Tyndall Air Force Base in Florida and at Army installations at the McGregor Range in New Mexico, then came on in to San Diego."

While in San Diego their host, Bob Schwanhausser, arranged a screening of the company's humorous movie, "Nobody's Perfect," showing some of the most laughable, though occasionally serious, test flights which went berserk. In addition to chuckles all around, the willingness to show such incidents (they occur in every company's experience) brought this positive reaction – "It was great. It sold us, because any company which can laugh at its own mistakes, has obviously fixed them well!"

"The Israeli team," Schwanhausser continued, "again headed by Uri, returned to San Diego in August. This time he was accompanied by Samuel Dror, their New York purchasing agent. In the period of a day we scratched out the specifications, a delivery schedule and a price.

"They wanted to finish work by six o'clock, but at 5:30 we still had not reached an agreement on price. Uri made a beautiful speech as to why the price was too much; that he couldn't recommend buying the drones. I said that we only negotiated once and that we had only a few minutes.

REVIEW BOARD INSPECTS FIREBEE power plant *at Army's McGregor Range. From left to right: Uri Talmor, David Klein (Cyclone), Levi Tsoor (F-4 pilot), Yoash Tsiddon, and Jack Rathgeber (TRA Base Manager).*

Dror said there will be no further negotiations and looked straight at me for the next move.

"I retired with Frank Jameson, company president, and our financial people to caucus for about ten seconds. It was all very formal. Everybody was 'Mister.' We came up with a new dollar number and made our counter proposal. Then the Israelis broke out in broad smiles and said the proposal was accepted. We finished at ten minutes to six just ahead of the planned deadline and went out for a drink together. From then on it was 'Bob' and 'Bondi.'

"At that time we were still discussing what cameras to use and whether we could furnish them or they would buy them separately. So much time went over the dam in this process that we had to remind them delivery was going to slip if they couldn't finalized their decision.

"Soon General Benny Peled arrived in this country to visit all the optics companies that had camera proposals in for consideration. They bought a brand new camera for high altitude missions and modification of an existing system – a much better camera – for low-altitude.

"They ordered a dozen RPVs of what we consider to be the best low altitude remotely piloted vehicle we have ever built. It ended up almost like the concurrent 147SD version but the 124I will also do the high altitude mission. You change the windows in the nose and you change the cameras but it is the same airframe configuration – the basic 124 with some extension in fuselage length and wing span. Like the 147SD, the 'I' bird has the larger 1920-pounds-thrust CAE J69 engine and, unlike the SD, is ground-launchable as well as air-launchable.

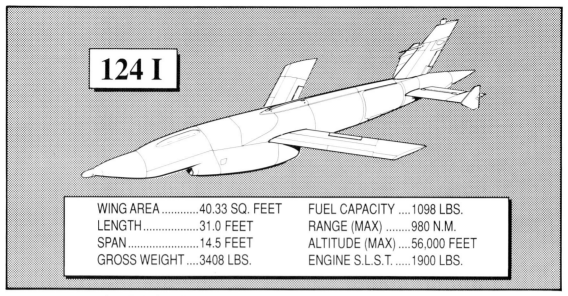

WING AREA	40.33 SQ. FEET	FUEL CAPACITY	1098 LBS.
LENGTH	31.0 FEET	RANGE (MAX)	980 N.M.
SPAN	14.5 FEET	ALTITUDE (MAX)	56,000 FEET
GROSS WEIGHT	3408 LBS.	ENGINE S.L.S.T.	1900 LBS.

ONLY THE NOMENCLATURE DIFFERED between the Israeli 124I and similar U.S. birds. State Department ruled out use of 124I for weapons suppression and tactical missions.

"OF COURSE WE HAD State Department approval for our visit to Tel Aviv in March, 1970 to present the 124 training target bird. We had to come back after the proposals were written by Cyclone to go through a licensing arrangement with the U.S. government. To make it easier for the people in government to handle, we maintained the 124 drone nomenclature of the BQM training target throughout, but Washington knew precisely what was going into the bird.

"By this time, Ray Ballweg was Vice President of Teledyne Ryan in charge of our Washington office. He had been in on reconnaissance drones from the very beginning and handled the paper work through the State Department. He presented the whole story and openly explained that we were using the 124I nomenclature but that it was indeed a military reconnaissance vehicle.

"Because reconnaissance drones, being unmanned, are considered nonpolitical – and being unarmed are not attack vehicles – the State Department considers them defensive in nature rather than offensive. Surveillance flights of an unmanned aircraft even over an area that is politically sensitive are still considered acceptable.

"For export, however, State would undoubtedly be against the sale of American drones for weapons suppression or other tactical type missions.

"Drones for reconnaissance are very attractive to the Israelis because of the tight control which has been placed over the number of American F-4 Phantom and A-4 Skyhawk aircraft they have been permitted to buy. If the drone could do the job, then they could release an F-4, for example, from a reconnaissance flight to do the attack or intercept type mission. They felt, too, it might be easier to get a license to import a drone and eventually to build them, and gain some independence in this manner.

"Sometime after Benny Peled returned to Tel Aviv he was promoted to Vice Chief of the Air Force. Uri Talmor, promoted to full colonel, moved up to wing commander. Since he had decided, while carrying the requirements responsibility, to buy the 124I reconnaissance drones, they gave him the job of operating them once the factory began deliveries. It's typical of the way they do business; that's one way they get the very real advantages of project continuity with their people."

FINAL CHECK-OUT at TRA factory in San Diego. Later IAF requirements resulted in some of same equipment for USAF drones.

Teledyne Ryan Aeronautical

Teledyne Ryan Aeronautical

RUN-IN OF 124I's JET ENGINE to check performance *preceded first deliveries to Israel.*

ONCE RYAN HAD THE CONTRACT in hand, engineers and production supervisors at the factory gave the new 124I program the old college try. "On the engineering side," said Pete Petrofsky, "Frank Oldfield, with responsibility for the flight control system, and Bill Wall, deserve a lot of the credit."

Robert Perkins, then the company's new home-base coordinator of international programs, pointed out that "While the projected 147SD reconnaissance vehicle and the new 124I were expected to have a lot in common, the 'I' actually preceded the SD out the door.

"Many of the design features that engineering was developing for the 'I' were later rolled back into the SD which was delivered after the first Israeli birds."

"The 'I' and SD were complementary types," Petrofsky pointed out, "but because of Israeli requirements, features went into the 124I that the SD did not have. However, the U.S. Air Force 147SD birds were later retrofitted with the same improvements. By contract terms the 'I' was built from scratch with no SD parts in it so that it did not impinge on the SD at all."

THE DATE OF THE FIRST demonstration flight in Israel – Monday, August 23, 1971 – had been set months before after lengthy scheduling conferences at the Ryan plant in San Diego.

Now, precisely on schedule, the first 124I Israeli bird was about to be launched in the desolate regions of the Sinai desert between Israel and Egypt. For Ryan it was critical because this particular model had never been flown – it had been sent halfway around the world for its first flight – and it was to be a ground launch, much more difficult than the usual air launch.

Frank Marshall, director of test and systems support and in charge of the Ryan demonstration group, later recounted highlights of his 2-1/2 week stay:

FIRST FLIGHT

The actual launch was made three minutes late, at 2:33 PM, after brief delays due to range camera problems including perhaps a minute due to a chase aircraft delay.

The bird was launched – the entire flight from launch, JATO bottle separation, climb out, flight programmer, performance on manual control – all were virtually flawless. The bird did exactly what we had expected it to do. The recovery sequence, the mid-air retrieval, docking and stowing of the bird, the return to the operational base – all were done without incident. In fact, done with a great amount of flair. The bird didn't even have a scratch.

The entire operation including preparations for launch, most of the flight – on a beautiful clear day – and the actual recovery sequence could be seen by everybody.

Many of the top military people of Israel were there – General Motty Hod, commander of the Israeli Air Force; Lt. General Haim Bar-Lev, chief of staff of the armed forces; Colonel Uri Talmor, wing commander; Colonel Cohen, base commander.

Following return of the bird, the recovery and photo helicopters parked and the crews off-loaded. The Israeli military people were called to attention. Colonel Talmor produced half a case of champagne and glasses – and proposed a toast "to the finest, most trustworthy commercial organization that this country and this Air Force has ever worked with; we salute and toast the Ryan Aeronautical Company."

There were moments, however, when we had some misgivings.

We were, of course, all experts in operating reconnaissance drones. This was to be the Israeli Air Force's first exercise.... and they are proud people. We found there was a tremendous atmosphere of professional competition between the Israeli forces and the Ryan crew; but it was really a pleasant thing to behold.

All along we had been assured not to worry about seemingly overlooked details. They would take proper care of everything and on time. They were very sensitive if we reminded them more than once about something that they had not done.

The first thing that concerned me just before the launch was that I didn't see the chase

AS REMOTE AS THE MOON, and secure from prying eyes, *the isolated IAF base 'somewhere in the Sinai' is new home for the 1241.*

pilot in his F-4 jet anywhere around the runway. He was due in the air five minutes before launch and that was only ten minutes away.

I said to Colonel 'Cheetah' Cohen, "Old friend, are you going to make it?" and he replied, "Yes sir, don't worry about that."

Just then, exactly five minutes before launch, out of some hole, somewhere in the distance, came this F-4 thundering down the runway and into the air right on schedule.

The pilot arrived back over the ground launcher at the scheduled time, but the JATO bottle had not been ignited to launch the bird, so he pulled back the stick and sent the F-4 into a vertical climb.

When the bottle lit off, he coolly rolled the plane over to acquire visual contact; then leveled out at about 500 feet to fly formation perhaps only 150 meters from the bird during its entire flight.

We did have one in-flight crisis which had its light moments. Ryan provided a back-up director system, called BUDS, which is essentially a portable microwave command and guidance system should something go awry with the MCGS unit. If you lose the MCGS carrier, the drone engine will shut down automatically, the recovery parachutes will come out and the bird will recover wherever it is. You hardly want that on your first mission, but it nearly happened when we lost the MCGS carrier.

Inside the MCGS van we had Ed Sly controlling the bird on its demonstration flight with Corrado (Pat) Pratarelli backing him up. Just outside the van was the portable BUDS with Bob Todd as emergency controller.

Even before the lost carrier, there was plenty of activity at the van. With all the VIPs there, the excitement was running as high as I have ever seen. It's a miracle that everything came off as well as it did.

THE MOST EXCITED OFF ALL, was Colonel Uri Talmor who had to keep the VIPs tuned in as to what was happening throughout the flight.

When the crisis came Pratarelli began rushing back and forth between Sly and Todd while control was transferred from the van to BUDS. At one point, Colonel Talmor got in the way and skinny Pratarelli put a football player's body block into him which sent him head over heels, but not a word was said. The Colonel understood we were in the middle of a crisis. He dusted himself off and went back to tell the Generals that things were moving along pretty well.

As far as the bird itself was concerned, it flew an excellent mission due to the skill of the Ryan controllers during the lost carrier incident. Toward the end of the flight I began to worry about the recovery sequence.

'BUDS' SAVED THE DAY on the first flight. *The Back-Up Director System and TRA's Bob Todd performed as required when the chips were down.*

The helicopters for the MARS mid-air retrieval were due to be airborne 15 minutes before scheduled recovery time. None were to be seen; no tell-tale signs of swirling dust from chopper blades were visible anywhere. Again I had a few words with Colonel Cohen.

"Old buddy, I don't see the helicopters around anywhere." "Don't worry, Frank, they'll be off right on time."

Sure enough, 15 minutes prior to recovery time, off in the distance, apparently out of some hole, came two H-53 helos, flying side by side in tight formation. In real precision manner, they positioned themselves at the proper altitude for recovery and at a location where everyone could observe the actual recovery operation.

One chopper neatly snagged the parachuting bird while the other supplied close-in photo coverage of the action. It was a sight to behold. The showmanship was superb as the recovery helicopter flew back and forth over the runway continuing the reeling-in process so as to always be within sight of the spectators.

When reeled snugly in, the bird was gently lowered in front of the launcher; the load line dropped and the helicopter flown away to a very precise parking next to the camera helo. It was a slick and very professional demonstration of great proficiency on the part of Israeli Air Force personnel.

The bird had flown just two minutes short of an hour on its shakedown demonstration flight and had covered some 350 n.m. The MARS recovery had been well executed and there was no damage to the RPV (remotely piloted vehicle). The chase pilot reported 'an extremely stable RPV throughout the flight.' It was a good beginning.

MARSHALL'S PLAN

ONCE IN THE COUNTRY [as Marshall's report continued], the Ryan crew swung into a fastpaced daily routine.

The Dan Hotel in Tel Aviv was headquarters. Since the team had arrived in Israel during the peak of the tourist season, eight of us were rooming in Room 501, the Presidential Suite of the hotel. To the delight of the housekeepers and maids this became known as "Kibbutz 501" and the service received from then on was magnificent.

At 6:45 each morning a military bus came by to pick up our civilian team and transport us to Dov, a small airport on the outskirts of Tel Aviv. There we boarded a French Nord transport, similar to the U.S. Flying Boxcar,

and were flown by the Israelis an hour south to an isolated Air Force base in the Sinai desert.

While there, we are really out of communication with everybody; there is no way for us to talk to anyone in Tel Aviv or anywhere else. Though nothing was said officially, it is mutually understood that we are not to be seen at this location. It's clear that it's equally important that we not see what else goes on. The precautions taken made that obvious.

Our work area at the maintenance hangar is isolated and the only time we leave is when the trucks come to take us to the mess hall. By that time everyone else has been cleared out, so we eat alone and are not seen.

OPEN HANGAR FOR FIREBEE maintenance operations *provided shaded work area in hot desert.*

When the day is over at 5 o'clock the plane is waiting, everybody climbs on board and we fly back to Tel Aviv.

When the Hercules commercial airfreight from the States arrived with the birds and hardware, the plane was crammed with boxes and equipment. The Israelis were not fully prepared to receive visitors and our classified project so the Ryan types faced a huge task to meet the scheduled launch date.

As the boxes were unloaded out came the usual screwdrivers, hammers and crowbars to remove the lids. That would be too slow for the likes of Bruce O'Hara, a well trained and qualified samata expert. Similar to karate, samata is the Thai art of foot fighting.

O'Hara probably doesn't top 5'8" nor weigh more than 140 pounds. "Just get the hell out of my way," he says, "I'll take those lids off."

With almost the 'speed of sound' he went from box to box, leaping into the air and with the heels of his heavy shoes he had those lids flying off the boxes like you never saw before. With one kick he sent an extra large lid sailing. Nearly eight feet square, it hit Bill Forehand in the chest hard enough, it seemed, to break every bone in his body, but he merely fell back and recovered his balance.

AS SCHEDULED LAUNCH DATE neared there was so much to be done the Ryan crew on several occasions felt they

Ryan Aeronautical Library

ISRAEL's SQUADRON COMMANDER *Major Shlomo Nir and family share R&R with TRA's Pat Praterelli, left, and Bob Todd, right.*

would have to work through the night. They took old shipping crates, made improvised mattresses of packing materials and ended up with a communal bed which served to give them a few hours' sleep at night between each man's specialized task.

Nobody asked them to stay there and work through the night. Guys like Tom Nietz, Gordon Allen, Jeff Grady, Tony Kaminski, and Joe Doering, made the decisions for themselves. I would have preferred they go back to Tel Aviv for a good meal and a night's sleep at the hotel, but their interest and dedication was so high they would rather stay at the site and get the job done to their own satisfaction.

Our people had a hard time getting adjusted to the food and water with the result that virtually everyone got sick at least twice during the first few weeks. The fever, chills, stomach cramps and diarrhea usually lasted from two to four days. It was no time to be working out in the boondocks and heat of the Sinai desert but some of them wouldn't stay in bed for several days on the recommended hot tea and bread regimen. They felt they had to go to the operating site regardless.

The amenities there were minimal to say the least. The outdoor latrine of corrugated metal was up the side of a huge pile of dirt. There was absolutely no ventilation, the daytime temperature inside was near 140 degrees and the stench made it oppressive to the point people just refused to go in there.

But when you had a dire diarrhea emergency it was a pretty sporty course you had to run.

As an old ex-field service man I decided things had to be improved and came out with my own version of the Marshall plan for a latrine that would be ventilated, odor free, and near at hand for any emergency. The gang set to work to build it, and next day in Tel Aviv I set out to give it the final American homey touch – a toilet seat.

MY QUEST FOR THE TOILET SEAT ended up as an hilarious experience. I couldn't speak Hebrew and the local people couldn't understand English. Finally, in desperation I walked the streets listening for anyone who looked Jewish and was speaking English. Finally, of all people, two young girls, maybe 17 years old, appeared to have the necessary qualifications.

I asked them where I could find a hardware store, and they said they couldn't recall having ever seen one, but they agreed to ask someone.

Finally they stopped an elderly Jewish couple and learned that "we don't have stores like that," but "what did you want?"

I said, "You're not going to believe this, young lady, but I am trying to buy a toilet seat." After expected and suitable snickers they got hold of themselves and passed the message along to the Jewish couple who joined in the merriment with the further question, "What do you want one for?"

Well, because of security, I wouldn't explain why I really needed a toilet seat in the middle of the Sinai desert, so settled for the explanation that I wanted to have one with me just in case I needed it!

I finally got directed to a plumbing shop down the street where only Hebrew was spoken. There I found the much needed article, selected a suitable one and finally convinced the proprietor to take a suitable amount of money from my wallet and let me proceed on my way with the precious toilet seat.

Back at the site next day, it was suitably installed in the 'chic sale' above the freshly dug trench and put into service. Only trouble was when the first man, Paul Carpenter the 'test pilot,' sat on it it slipped off and fell into the trench. After this initial failure the new system proved highly efficient and much appreciated by all clients.

ALL THE COMFORTS OF HOME were available *when Frank Marshall applied his know-how to this modern facility.*

AFTER THE INITIAL SHAKEDOWN flight of August 23rd the Ryan and Israeli crews swung into a full schedule of demonstration flights before final acceptance of the dozen birds. Meantime, Marshall had completed his rather brief assignment and was replaced as on-site program rep in September by another experienced Ryan pro – Pete Petrofsky.

He was there for the fifth demonstration flight, September 24. It was to be one of the final checks of all systems with particular emphasis on the low-altitude Radar Altimeter Control Equipment (RACE), the Doppler navigation capability and the camera systems for low and high-altitude photography.

"On this, the first low level flight," Pete explained, "we were programmed to fly at 300 feet but in passing over some of the hills in the Sinai desert it became erratic, the bird beginning to porpoise excessively through plus or minus 50 feet."

The erratic flight caused the on-board electronic sensors to believe there was a marked change in the terrain below and activated the TEP system – terrain evade procedure – which brings the bird sharply up in a 7g. climbing maneuver for terrain avoidance. Back at 7000 feet altitude the TEP was programmed to return the bird to the 300 foot level, but it failed to respond rapidly enough and according to the chase pilot the bird dove into the sea on a northerly course.

"That," says Pete, "created problems not only in Israel but also back in engineering at San Diego. Three of Teledyne Ryan's top management team, Bill Rutherford, president Frank Jameson and Ray Ballweg came over and I stayed there until we all returned to San Diego.

"Working day and night, engineering came up with some good redesign to solve the low-level flight problem. Then I went back to join the technical team in Israel.

"By then we were confronted with a real weather problem with low clouds over the Sinai for several weeks. We spent a lot of time running back and forth between the desert test site and our hotel in Tel Aviv.

"Col. Uri Talmor was running the show and wanted to fly the low altitude test flight in conjunction with using the camera. We had already tested the camera system and knew it worked perfectly. My position was that 'we have already done that Uri, so let's just do the low-level test whenever the cloud cover will permit.'

"Uri finally consented and that flight went flawlessly.

"The bird flew at the prescribed altitude of 300 feet, right down the main runway of the Bu-Ga-Gafa airbase and flew off into the countryside. The RACE capability had checked out okay. But there's an unfortunate side to that flight.

"ON COMPLETING THAT RUN, which was the primary purpose of the flight, the bird was brought to a higher altitude in sight of Bu-Ga-Gafa and put into a parachute.

"It was caught by a CH-53 helicopter – a real nice catch – and while they were reeling it in the reel started to slip and instead of closing in on the helicopter the line was further extended. The pilot went into a crash maneuver trying to save the bird but couldn't do it.

"The cable was stripped from the drum and the bird rotated to a vertical nose-down attitude and crashed into the desert within a quarter to a half mile of a tank brigade on maneuvers. They thought it was some sort of a missile and closed in on it but by then another helicopter had landed and shooed them off.

HIGH OR LOW FROM BU-GA-GAFA, the 124I *demonstrated it could really RACE with its Radar Altimeter Control Equipment.*

Israel Aviation & Space Magazine

NOT A FLYING TIGER but a flying shark *ready to find the launch site of enemy SAMs.*

Jerusalem Post

"The Israeli officer who I was with at the control station was a helicopter pilot and he had an Alouette helicopter there – so he grabbed me and we flew out to where the bird was. It was ironic that the bird picked up speed as it fell, so it started to fly.

"The nose was coming up and rather than impacting at 90 degrees it probably impacted at a shallow angle of about 30 degrees. It almost made it but instead ended up spread all over the place."

Everyone, including the IAF, agreed that the flight had been successful since it was the reel system in the retrieval helicopter, and not the 124I, which was at fault.

ISRAELI OPERATIONS

SOMEWHERE BETWEEN the Ryan-conducted demonstration flights of September 14 and 24, the Israelis ran a mission of their own. Reportedly an operational flight was made over SAM sites in Egypt in an attempt to scramble some MiGs and trick them into chasing the drone over Israeli held territory. That ploy didn't work, but the bird did draw several surface-to-air missiles, none of which found their mark.

In connection with their own missions, the IAF has offered some comments on personnel and operations:

"The [remote control] operators are aircrew personnel. There is no requirement to be a pilot. . . but in order to integrate into the IAF operational system, we need people who know the system well. Some operators are engineers as well as pilots, which contributes largely to the expansion of the RPV's operational applications.

"Here the technical team is fully engaged in all the details of the mission. It is essential that technicians always participate in operational briefings since they program the mission plan into the RPV systems, and check [the bird's] airworthiness prior to launch.

"Master Sergeant Yehuda, who supervises the electronic test consoles, had joined the first group sent to TRA for the Firebee procurement, which formed the operational team for the premier operational flight in Israel.

"'We tested the aircraft using these consoles, at the launch site, programing the mission data,' recalls Yehuda. 'The Firebee flew along the Suez Canal line and after an hour or so came back. Its parachute opened and the helicopter brought it down. Immediately we checked it and started to prepare it for the next mission.'"

ISRAELI INTELLIGENCE plays its cards pretty close to the vest. Once the 124I demonstration flights were completed (the last one a high-altitude test December 17), Ryan personnel – though regularly sought out for technical assistance – were kept at arm's length when it came to operational missions.

The Israelis, of course, wanted to stand on their own two feet just as quickly as possible. They didn't want excess onlookers around not only from the standpoint of security but it could also put contractor personnel in the embarrassing position of being party to or knowledgeable about their operations if they went beyond strictly reconnaissance missions.

"They felt very strongly about this," said Bob Schwanhausser. "That's one of the reasons when an F-4 Phantom was flown into Israel by an American pilot, he had only 24 hours to leave the country after the delivery was made. All of the Teledyne Ryan drone equipment was flown into Israel by commercial air carriers, not military.

"These people turned out to be the most hospitable with whom we had ever worked – and the most dynamic as far as decision making is concerned.

"Defense Minister Moshe Dayan, hero of the 1967 Six-Day War, was on hand for our first flights as were General Haim Bar-Lev, chief of staff, and General Motty Hod, air force chief.

"The top brass get personally involved wherever possible. By our third demonstration flight, they were so pleased they decided it was no longer necessary to have the drone accompanied by a chase plane. That didn't sit well with General Hod because he had expected to fly the F-4 chase plane on the next mission, with one of the top procurement officials riding in the back seat. That's the type of overlook they give a project.

"The hospitality of the Israelis left nothing to be desired. In the summer of 1971 they suggested that I send over my 13-year-old son, Robert H. to live in a kibbutz. It was his first trip away from home or abroad and because of his age he was too young to go into the kibbutz so he was 'adopted' by Abraham and Phyllis Portugali. He was a sculptor who spoke little English and she spoke with a British accent but was having difficulty with Hebrew. Life on the kibbutz near Caesarea, between Tel Aviv and Haifa, was quite an experience for young Schwanhausser.

"Bernie Paul, who had accompanied me on our first trip, is Jewish and very proud of his ancestry. In a conflict of loyalties, Bernie had some tough choices to make because the Israelis urged him to immigrate because 'this really is your homeland.' However, he decided that the United States was his home and his country and that this did not in any way minimize his Jewish background."

DETAILED INFORMATION on Israeli operational reconnaissance flights during Middle East wartime conditions has always been hard to obtain. Press reports, however, often provided a general record since an enemy country generally seems willing to expose its opponent's tactics.

Thus in May 1972 it was reported that a 124I on a high-altitude reconnaissance mission overflew Cairo, returning to a recovery area in the Israeli-held Sinai Peninsula near which it had earlier been ground launched. Presumably its mission was to photograph Egyptian gun and missile emplacements west of the Suez Canal.

Apparently no effort was made by Egyptian or Russian forces to intercept or down the unmanned aircraft. However, Israel was earlier reported to have lost two 124Is since the birds went on operational status.

Several newspapers in the Middle East reported in October 1973 that Egyptian forces had shot down a Firebee-type drone over the Suez Canal but details were sparse.

Finally, in December, the Israeli government confirmed that it was using unmanned aircraft when it admitted the truth of "an Egyptian statement that one of its planes crashed (December 13) near the Suez Canal."

Cairo said it was shot down; Tel Aviv that it had spun out of control after deviating from its patrol routine.

Two weeks later both Egyptian and Israeli officials reported the downing of a pilotless Israeli reconnaissance plane over the Suez Canal by Egyptian missiles.

The Cairo communique reported: "The plane was seen crashing in flames on the eastern bank of Lake Timsa, midway along the 103-mile canal."

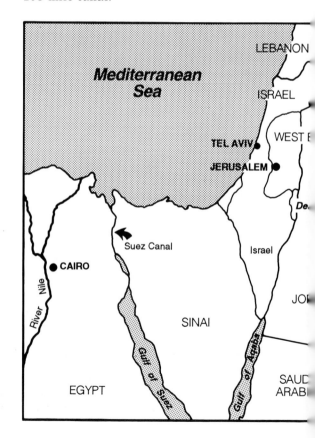

UNDOUBTEDLY THE MOST RELI-ABLE information available to the public came from the Israel Air Force's own official journal, "Heil Ha'avir," published in 1974 and available in book shops in Tel Aviv and other cities.

The journal, in a well-illustrated article, disclosed that pilotless aircraft carried out numerous missions in the October 1973 Yom Kippur War as an integral part of the Air Force. Decision to buy Firebee-type RPVs had been made in response to the war of attrition on the Suez Canal in 1969-70.

The drones were used over both Egyptian and Syrian territory, and the journal reported, brought back valuable photos of airfields and military installations. Many of the RPVs survived heavy barrages of missiles and attacks by enemy planes.

Excerpts from 'Heil Ha'avir' follow:

UP TO THE YOM KIPPUR WAR, the RPVs performed photography missions, commonly along the [Suez] Canal line. The missile defended zone was hazardous to manned IAF aircraft flying reconnaissance missions. Simultaneously, the RPV unit carried out training flights and test flights for system performance improvement.

Immediately prior to the Yom Kippur War, the unit was set on alert and began preparation for operations. On the eve of the war's outburst, the unit was instructed to deploy in the north. The squadron, after having deployed to an isolated open field site, waited for the inevitable. At 1400 hours, they saw MiG's passing over them, and explosions.

Starting at first light, the unit began to launch RPVs, still awed by the spectacular sight. Not

CONTROLLING FLIGHT OPERATIONS from a sheltered bunker are Lieut. Aharon Shohet, left, and the 200th Drone Squadron's Commander, Major Shlomo Nir.

many hours passed, until the unit received mission orders for the Egyptian arena as well. While ready for launch, they were attacked by two Egyptian MIG-17 foursomes. The personnel were not injured; they shook the dust from their protective vests and launched the RPV's on time.

At the same time, we received an order to operate the Firebee for intelligence purposes. The teams operating this vehicle were deployed to the appropriate launching sites. The unit operated both RPV models [The IAF also used the smaller Northrop Chucker drones – Ed.] from many launching points. The number of missions increased considerably as the squadron personnel decreased the preparation cycle period. Since heavy helicopters were busy with priority operational missions, the RPVs could not enjoy

FIRST PHOTOS OF ISRAEL's RPVs flown in Yom Kippur War appeared in "Heil Ha'avir," the Air Force Journal, in 1974.

the mid-air-retrieval system, at least in the beginning of the Yom Kippur War.

Cairo's citizens took shelter several times during the war. Now, we are free to reveal, that a considerable amount of these alerts were caused by the Israeli RPVs passing above the city on their missions.

Several days after the outburst of the Yom Kippur war, the RPV squadron received an order to perform a 'base survey' above Egypt for reconnaissance purposes. The unit commander arrived at the IAF HQ with his plan consisting of a long snaky route over airbases and many other installations. The RPV had to take many risks.

Immediately after launch the digital programmer failed, resulting in a choice between cancellation of the mission or its continuation by remote manual control. The RCO [remote control operator] decided to go on with the mission. The RPV crossed the Suez Canal flying towards Cairo.

The failed auto-photographing command caused the RCO to make a decision to try manual photography. He instructed the control station radar technician to send direct camera operation and stop commands to get the necessary overlapping covering the desired area. Simultaneously he assumed remote guidance of the planned route, following the plotter map display. The failure called also for a decision to slightly change the route, approaching Alexandria.

BENEATH AN IAF H-53 HELICOPTER, the drone, *with its directional drag chute trailing, is returned safely after a recce flight.*

ON THE WAY BACK the mission also suffered a loss of radar communication with the RPV. The commander believed that the RPV was shot down, but the IAF radar control reported that it was still flying, chased by MiGs and missiles.

The RPV continued its route however, keeping the last heading and altitude prior to the loss of communication. The MiGs did not succeed and close to the canal crossing the RPV was redetected and communication re-established.

The RPV landed safely, after a flight of more than an hour and a half covering hundreds of miles, bringing home a variety of photographs – Egyptian airbases, missiles, army concentrations and urban views.

Says the RPV unit commander, "This flight proved the RPV's potential performance and directly caused a positive shift in the confidence given to this new vehicle."

Above Syria as well, the RPV's performed 'base surveys' providing photographs of military and civilian installations. A MiG-21 shot down a RPV whose debris was photographed and published in the Parisian newspaper 'Paris-Match' on Oct. 27, 1973.

In another incident, a RPV conducted a "blind dog fight" with Egyptian G/A missiles. Each time a missile was launched at the RPV, the RCO maneuvered it sharply. Only after launching many missiles the Egyptians managed to shoot one down. When the Egyptians found its debris, they surely enjoyed the funny drawings and accompanying inscription painted on it especially for this mission.

An RCO returning his RPV for recovery, has the habit to inform his control tower: "I am coming to land" as if he was inside the vehicle. During the war, the number of missions increased to a rate where a control tower flight controller approached the RPV with landing instructions as if it was a manned aircraft. Seeing the recovery parachutes, he almost reported pilot bail out.

"WE UNDERSTAND," says Petrofsky, "that in the Yom Kippur War the 124Is were used primarily over Egypt both as reconnaissance vehicles and in some cases as decoys so they would be mistaken for manned combat aircraft.

"In the latter role they would draw enemy fire. Reportedly all the SAM surface-to-air missiles at one site were launched against the RPVs. Then the Israeli Air Force could go in with fighter-bombers and wipe out the whole enemy installation."

During the closing phases of the Yom Kippur War, an Israeli delegation again vis-

SUPPLEMENTING ITS 124I RECCE BIRDS, such as above, Israel has ordered Model 232 training targets equipped with Microprocessor Flight Control Systems (MFCS).

ited San Diego to see if TRA could produce a recce version of the recently introduced supersonic Firebee II.

"We did some preliminary design work," Bob Perkins reported, "on a stretched bird using the same KA-93 camera they had in the 124I. They also sent over their best aerodynamicist to work with our people.

"Such a modified Firebee II would have been able to launch from Israel, fly out over the Mediterranean, come in north of Lebanon into Syria to get all the hot targets in Syria, and still make it back into Israel for recovery – and have all of the over-land portion flown at supersonic speeds.

"They felt they needed the higher-performance capability as soon as possible but finally concluded that fully developing a new supersonic recce capability was more time consuming than just ordering more 124Is."

By this time it was reported that the IAF was down to only a few birds. "After the Yom Kippur War," Petrofsky says, "and as a result of a good relationship with Benny Peled, he arranged a briefing with his staff to review – yes or no – the value of the 124Is during the war. The answer was a very positive 'Yes!'

"It was a result of that type of feeling that within several months they ordered another batch of 124Is."

IN THE EARLY '80s the Israelis saw a new opportunity to modernize their reconnaissance capability.

"The U.S. Air Force," says Bob Perkins, "decided to clear its inventory of Vietnam era 147Ts and 147TFs in storage and transferred 33 to the U.S. Navy with the idea they could be converted back to aerial targets, but the Navy's plans did not materialize.

"The IAF came over to look at them as they were stored between manufacturing buildings at TRA's San Diego plant. Israel decided to buy them as surplus from the U.S. Navy.

"These birds had engines, flight control and were basically complete and in pretty good condition. However, scoring and other classified payloads were not included.

"Israeli thinking was to update the birds to later configuration, retaining the basic airframe and adding a capability for land recovery. However, the l47T series were all air launched and these thin-skinned birds couldn't take the landing impact very well.

[Although the IAF is said to have wanted an air-launch capability it was thought to be politically unacceptable because all operations should originate within the confines of Israel as a point of launch. – Ed.]

"Plans for TRA to do the necessary updating fell through when the Israelis decided, some time later, to spend funds when they became available, on a new configuration.

"As far as we know the 147T and TFs are still sitting somewhere in a nice dry area in Israel in the sea-going 40-foot containers in which they were shipped from San Diego.

"In the interval we had shipped the IAF their second batch of 124Is. Today, 1990, they still have an operational capability of some portion of that order.

"It's not just because we built one hell of a fine aircraft for them but that they have done an outstanding job of being able to maintain and operate a system which, by today's high-tech standard, is antiquated.

What they use – bailing wire, chewing gum – I don't know but they may go up with an operation fully aware that somewhere in that flight a critical item may fail – so they just plan around it.

"Some missions may start out as remotely controlled and wind up being manually controlled to get it back but they get the mission done.

"That kind of performance is a real credit to the Israeli Air Force."

TWENTY YEARS AFTER introduction of the 124I to Israeli skies, the inventory of flyable RPVs had been reduced to a handfull.

The problems in maintaining the photo reconnaissance capability of the IAF was highlighted in the April 1990 issue of the Israel Air Force Command's magazine.

It was pointed out that the RPV squadron had received a letter of appreciation from the Air Force Commander which was logically interpreted by the squadron as more than just a thank-you letter. For them "it was a renewed trust in the urgent need for the squadron's existence."

It is axiomatic that only four months after appearance of that article new tensions broke out in the Middle East. Iraq had just invaded Kuwait!

ISRAEL'S INTEREST in the Ryan Firebee began with its use as a reconnaissance vehicle, but following the Yom Kippur War of 1973, the need arose for a vehicle capable of testing the effectiveness of new missile systems then being developed in country.

This meant having new pilotless target aircraft (PTAs) for the necessary evaluation flights.

"Israel's initial 1974 order for five Firebee vehicles was not for the standard versions of our training targets," reports Bob Perkins. "They wanted performance that included higher G manuevers, requiring a more sophisticated flight control system than the two axis system used in the U.S. targets.

"That requirement became the genesis of the type of three axis control system we have today. We modified the standard two axis system by adding a third axis for rudder control.

"From the first flight that winter the newly configured vehicle performed as advertised. Further deliveries to Israel of the **Model 232** version of the Firebee with its hybrid flight control system were made in 1977 and 1978.

"New orders for the 232B were placed in 1982, this time for a further updated flight control system. These vehicles became the first Firebees to be equipped with Teledyne Ryan's new Microprocessor Flight Control System (MFCS). This system was subsequently adopted by the U.S. Air Force and Army for their new production Firebees.

"While Israel is not a large user of these Firebees they are very capable operators who get the most out of these target systems. They have since placed additional orders in 1985 and 1990."

JAPAN

Photo by STEVE UYEHARA

SHIPBOARD LAUNCH FROM 'AZUMA'

JAPAN: Targets for Defense

IN THE MID-1960s as the war in Vietnam begin heating up it caught the attention of the Japanese because of the effect it would have throughout Asia coupled with the ever-present concern of a threat from Red China.

The wisdom in Tokyo was to further hone the skills of the Japanese Self-Defense Forces.

THE JAPANESE MARITIME Self-Defense Force was somewhat limited by restrictions imposed in the final treaty signed by the Japanese and the U.S. as to the types and numbers of defensive weapons they were allowed to manufacture in their home country.

However, building unmanned aerial targets for missile proficiency for their Navy ships was not on the restricted list. Consequently, a search of qualified U.S. target manufacturers was conducted with Ryan Aeronautical finally chosen to enter into a period of negotiations leading ultimately to the manufacture of **Model 124** Firebee Q-2C drones in Japan.

JAPANESE MARITIME SELF-DEFENSE OFFICERS were eager students *during plant visits and briefings conducted by Doc Sloan.*

Ryan Aeronautical Company

Fuji Heavy Industries, one of the great manufacturing complexes arising from the destruction of the war, was involved in producing many products to rebuild the country.

Fuji's large plant in Utsunomiya, a hundred miles from Mt. Fuji, had been the designer and builder of WWI 'Zero' fighter planes. Later, essential production items included heavy railroad equipment such as freight cars and locomotives. Fuji was also a major subcontractor for Bell 'Huey' helicopter rotors and fuselages.

The trading firm of Nissho-Iwai in Los Angeles was designated as the U.S negotiator for Fuji. Maki Morino, Japanese-born but educated in Los Angeles, met in 1967 with Elmer Stone, Ryan's general counsel, to discuss the possibility of eventual manufacture of Firebees at Fuji. Joining them was Akimichi Kasahara from Fuji's aircraft department, Tokyo.

A contract was finally signed for **MQM-34D** type Firebees, resulting in a disclosure fee, major fabricated sub-assemblies and reproduction of engineering data. Also included was indoctrination of Fuji technicians in all aspects of manufacturing, tooling and assembly, as well as training in field operations.

Seven qualified technicians under the supervision of Hideo Takayama, spent two months working with their counterparts, not only at Ryan's San Diego facility, but off-site at White Sands in New Mexico. The writer (Doc Sloan), prematurely grey, was the Ryan coordinator for the operation, and the Japanese representatives were all rather short in stature. Leading his troops through the plant, the remark was often heard, "There goes Snow White and his seven dwarfs!"

In addition to the Japanese Maritime Self-Defense Force, their Ground Self-Defense Force (GSDF) was one of the driving factors in the procurement of Firebees. Their requirement was to conduct annual service practice against Firebees with Hawk missiles in Japan rather than sending troops for training against Firebee targets at the U.S. Army's McGregor Range in New Mexico, as previously done. The GSDF was the first Japanese group to purchase Firebee/Towbee systems for use as targets for Hawk Annual Service Practice.

The Ground Force purchased targets and accompanying support equipment without having a missile firing range in the home islands. They had actually arranged to buy the necessary land in Amori Prefecture for this application, but when radical leftists got wind that farmers were to be displaced with implements of war, there were mass demonstrations. Ultimately, the leftists prevailed.

Other sites that were considered were Sheznai (on the island of Hokido), Okinawa and Iwo Jima. The former site, which is today the principal GSDF gunnery range, was rejected because it was in the middle of race horse breeding country and the noise of the Firebee's JATO would have been sufficient to disturb the young colts. Iwo Jima also fell by the wayside. GSDF never did get their range, and in 1975 the Firebees were officially transferred to the Navy. To date the Japanese Army still fires their Hawk and Patriot missiles in training at McGregor Range!

SUCCESS OF THE FUJI/RYAN meetings was due, in a large part, to the relationship between Elmer Stone and Maki Morino. Elmer had spent several years in the Orient and was fluent in Japanese. He understood the Japanese procurement psychology thoroughly and, with his understanding of the necessity of rebuilding the country, was able to conclude a mutually satisfactory agreement.

During this period, the U.S. 'Tartar' missile was made available to the Maritime JSDF for their ships, but crowded shipping lanes in the vicinity of the home islands required off-shore operations well beyond the Japanese coastline.

Ryan Aeronautical Library

RAIL-LAUNCH FROM FAN TAIL DECK of the Azuma in a rolling sea *created problems eventually solved by TRA's Larry Emison and Hideo Takayama.*

Sea launch of Firebee targets at this time was relatively untried. The U.S. Navy had experimented at Pt. Mugu using a converted air-sea-rescue boat for sea launch, and later at Roosevelt Roads utilized a decommissioned destroyer under tow for close in-shore launch of Firebees.

The Japanese took a more creative approach, starting with building a ship designed solely for the purpose of conducting off-shore launch and recovery of targets on the high seas.

The Maizuri Heavy Industries shipyard in Kyoto was designated as builder of the auxiliary training ship 'Azuma' and construction of the ATS started in 1967. Azuma, with the ability to carry five flight-ready Firebees, was ready for sea trials by 1969.

Designed to rail-launch targets from a platform on the aft section of the weather deck area, the Azuma III contained complete facilities for turn-around, recovery, decontamina-

Japanese Self-Defense Forces

DESIGNED AND BUILT SOLELY as an aerial target operating facility, *the Azuma steams from Kyoto out to sea for her initial trials.*

tion and maintenance. Control stations, both direct and remote were incorporated as part of the package. The 2000-ton ship measured 320 feet in length, carried a crew of 180 officers and men, with a speed range of up to 18 knots.

THE NAME AZUMA has a history dating back as far as the third century. Retired General Masumoto of Nissho-Iwai's Tokyo office described the origin of the name in a letter forwarded to Ryan's Bud Miller in 1969.

"By Japanese dictionary," he explained, "Azuma has a binary meaning; in English the words are 'east' and 'my wife.' The origin of the name comes from an incident in the life of Prince Yamato-Takeru following the suicide of the prince-warrior's wife.

"The first ship to bear the name Azuma was a U.S. civil war vessel sold to Japan. The second Azuma, a French-built armoured cruiser, served in the Russo-Japanese war of 1904-5."

Larry Emison, Ryan's jack of all target trades, assisted in the design of the ship and worked closely with Hideo Takayama, program manager for Fuji. Prior to the activation of the ship, these two gentlemen spent a period of two years handling correspondence posing technical questions on both the targets and the ship. Loading the Firebee target on the zero-length rails in a rolling sea created a problem, but an innovative hydraulic crane designed by Hideo solved the problem.

C.D. (Bud) Miller took over program management for Ryan in 1968, and as project engineer, made several trips to Japan to brief the Maritime JSDF personnel on all operational aspects of the system.

In a lighter vein, during these briefings, a

WITH FINGERS CROSSED and facing as yet unknown situations, *the first Japanese Firebee is prepared for the critical System Qualification Test*

Steve Uyehara

pronunciation idiosyncrasy brought smiles to the serious tone of the discussions. It was during one of these meetings that Bud, always articulate and using his hands to emphasize a point, cut his hand on the glass edge of the briefing table. Unnoticed by him, he continued to expound when one of the Japanese officers interrupted, shouting "Mr. Mirror, you're breeding on the taber."

Miller, a veteran of Firebee operations, had been Ryan's Base Manager at the Army's Mc-Gregor range in New Mexico, at the Navy Puerto Rico operation and had directed off-site technicians at numerous installations. He was aboard the Azuma, accompanied by Ryan technicians Steve Uyehara, Ernie Moser and Ralph Sargent, for the initial test and evaluation flights. Here is his story:

"THE SUMMER OF 1970 was one of the hottest on record in Japan and when we all arrived at Ita Jima, the home of the 11th Fleet Training Group, the fun really started. For one thing, all of the Japanese sailors who had been through the Ryan training course now used a good deal of American slang intermixed with a fair amount of 'cuss words' and 'Americanisms' of their own language. For instance, Ernie Moser taught them the expression 'Jotie-Totie', which is a takeoff of the Japanese word Joto, meaning good, or great.

"In my travels around the night spots after hours many of the girls had on Ryan tie pins or were using Ryan cigarette lighters, which we had distributed to the training team in abundance to promote teamwork between the Ryan and Japanese technicians.

"Once the assembly and final checkout process was complete the final step in the formal acceptance process was the all important System Qualification Test, or SQT as everyone referred to it.

"This phase brought all elements of the program together for the first time as a true test of system performance. Success here really amounted to instant acceptance. A failure, on the other hand, could mean anything from project cancellation to long, painful delays.

"Somehow, all equipment and personnel were on board when the ship sailed for the SQT, although there were times when we were not sure that we would ever reach this point! One such incident occurred when the Japanese Project Officer announced that the eight U.S. manufactured JATO bottles were 'no good.'

"Apparently during the transit from the U.S. Navy's Indian Head arsenal to Japan, a process that took some nine months, the propellant had cracked in all but two bottles. If

PRE-LAUNCH PRAYERS IN THE ONBOARD Shinto Shrine *by the combined Japanese/Ryan crews were held to help assure the success of the initial flight. C.D. (Bud) Miller beams from second row center.*

Steve Uyehara

we had tried to use them the JATO would probably have exploded, conceivably killing someone or at the very least destroying a lot of expensive equipment.

"Fortunately, the Japanese had designed their own JATO unit from the U.S. counterpart, although it had never been tried before.

"Since the SQT required at least three successful flights, here would be another first.

"As I recall, everything associated with this test became more and more difficult as the appointed D-Day approached. For instance, we had trouble getting the targets on and off the launcher (the famous Fuji Jib Crane did not work properly); the EPSCO track and control system was in small pieces until the day before the first launch; the supply people had ordered the wrong salt water displacement compound for engine decontamination; and we were unsure of our weight and balance because of the ship's motion in the moderate sea states that were being experienced.

"A further problem also existed because the Army's marginal EPSCO system was being incorporated for the first time on a ship. Since to this point it did not work very well on land everyone was worried. And because there was no real way to test everything as a system before it was delivered, we all started to lose confidence. Of course, we did not want

to convey these feelings to the Japanese although it did not make much difference as they were worried on general principles anyway!

"Almost like magic, when we finally got around to actually flying the targets, EPSCO worked flawlessly. To this day it can't be explained as the Army continued to have trouble until they traded them in for new Vega systems.

"To make matters worse, the captain of the ship insisted upon feeding us American style food cooked by Japanese cooks. . . it was really grim looking and tasting. And, as we had a few moments of rest between crises periods, by relaxing on the ship's fantail, all I could think of were old John Wayne war movies where the 'Japs' were the bad guys. Incidently, the Japanese flags are the same today as they used in WWII; so, the visual effects were very realistic. To further complicate our mental health a Shinto shrine was erected in the passageway between our rooms. When I asked the Captain about it he smiled and suggested we use it as a help in bringing success to the operation. We did!

"The actual SQT was almost anti-climatic when compared to the preparation phase. The Firebees all flew as advertised. The EPSCO worked flawlessly. The Japanese

RIDE 'EM COWBOY, SAN! **With a hoist for a lariat** *and wet suit for chaps, a Japanese diver is ready to herd 'ole Firebee into the Azuma OK Corral.*

JATO proved to be superior to the U.S. version in many ways. And except for one of the EPSCO guys coming down with a locally-acquired malady we all emerged unscathed.

"After splash-down, a tender was sent out to recover the chute and attach a pick-up hook to the fuselage. A Japanese sailor, equipped with a wet suit, rode the towed target back to the mother ship like a bronco-buster and was hoisted aboard with cheers from the entire crew.

"A total of three flights were conducted, which was enough to persuade the MSDF that they had a winner. And, many years later when I visited the first project officer (i.e., 'JATO's no good'), who was now a Captain and far more dignified than he was as a LTJG, he responded (with a smile) to one of my questions with a 'jotie-totie.' In retrospect, I think we were successful."

SHORTLY AFTER BECOMING operational in a normal mode, the requirement for a low altitude capability in the Firebee became urgent. Another TRA team under the supervision of Mike Savino and including Steve Uyehara flew to Japan with the current state-of-the-art low altitude kits and

completed their installation in early 1970.

Mike recalls, "Fortunately, our initial flights were text-book perfect. Following launch from the Azuma, the bird was vectored out about 100 miles, did a 180° and dropped from 7,000 feet down to fifty. Track was right on the button, and the target passed within visual range of the mother ship.

"The skipper of the Azuma (later promoted to Admiral in the Japanese MSDF) was jubilant. He also had a fondness for good Scotch. Since their naval regulations prohibited consuming any alcoholic beverages at sea, the skipper ordered a coast-bound course at full speed ahead at the completion of the last mission. The minute the fathometer showed reasonably shoal water, the anchor splashed over-board, and we all toasted the success of the low-altitude mission. And, we all did a thorough job of toasting!"

THE INITIAL LICENSING agreement and the continued cooperation between Fuji and TRA has paid cost-efficient dividends to the Maritime JSDF over the years, and given the Japanese further independence in maintaining their defense capabilities. Within a short period of time, Fuji had the capability

Steve Uyehara

WITH 'CATCH OF THE DAY' and ready for refurbishment *of the Firebee, and the entire crew, the Azuma heads for shoal water and a well-earned ration of spirits.*

of manufacturing the entire target, with the exception of the Continental J-69 engines, purchased directly from CAE. Flight control boxes were successfully duplicated, tooling improved and production in Japan continued to supply sufficient **BQM-34AJ** targets for 20 to 30 operations a year.

In early 1990, Fuji representatives again visited TRA for the purpose of outlining future cooperative ventures. Bob Perkins, TRA International Programs Manager, conducted a two-day seminar. Items covered included updating Japanese Firebees to the latest U.S. Navy configuration, utilizing Fuji's own three-axis Digital Flight Control System and additional engineering and field support in Japan.

The latest Japanese-built Firebee came off the production line of Fuji Heavy Industries

in March 1989, but under the most recent agreement, new configuration targets will be delivered for flight test starting in September 1992.

The updated Firebees will provide the Maritime Self Defense Forces with the ability to perform anti-ship missile simulation and mobile sea range missions, similar to those conducted by the U.S. Navy.

The JMSDF is also scheduled to receive a new training support ship that will replace the Azuma. It will be able to conduct dual Firebee target flights, enhancing realism of at-sea defense operations against multiple threats.

With more than 25 years of close association, the Fuji/TRA relationship continues to be an important chapter in the Firebee story.

NATO / SARDINIA

As a bird of prey, the hawk has a voracious appetite and kills with deadly accuracy. Its target can be smaller birds, rodents, rabbits or other small game. The Hawk surface-to-air missile system, named after this bird of prey, also pursues its prey with deadly accuracy, hunting for its favorite target, the Firebee, on ranges throughout the world.

Several western European countries, members of the North Atlantic Treaty Organization (NATO), had selected the Hawk system as one of their prime elements of defense against a low-level air threat. The Hawk system was being produced in the United States and the basic missile system was approved for export to NATO countries.

A European consortium, under license from Raytheon, entered into production of Hawk missiles needed to support the NATO batteries. For each lot of missiles produced, it was necessary that a portion of the lot be test fired to confirm the acceptability of the production run. As hundreds of the lethal rounds were to be produced by the consortium, a total of seven lot-acceptance test firings against different aerial targets would be required.

Testing the effectiveness of the missiles could not be accomplished over the densely populated areas of the continent, thus the open area of the Mediterranean Sea was chosen for the testing grounds. Under the code name HELIP (Hawk European Limited Improvement Program) the NATO Hawk Management Office (NHMO) in Paris selected the Italian Salto di Quirra Missile Range off the east coast of the island of Sardinia as the test site.

The Italian firm Aeronautica Macchi (Aermacchi), with General Mario Matacotta as General Manager, became the prime contractor for target support services. Aermacchi,

PREY FOR THE DEADLY HAWKS, Firebees are readied *for ground launch and flight over the Mediterranean's Salto di Quirra Range.*

Aeronautica Macchi

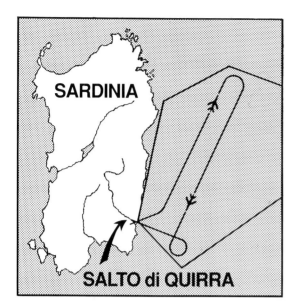

being an aircraft producer, had the personnel and managerial capability to oversee the proposed test operations.

With the range requirement settled, selection of a suitable subsonic-high maneuvering target was the next logical step. With the Firebee being used in that role against the Hawk in numeroud locations around the globe, the choice was an easy one. NHMO contracted with Aermacchi to buy and operate the targets, and Aermacchi in turn contracted with Teledyne Ryan to produce the **Model 232A** targets and ground support equipment needed. Also, TRA would train Aermacchi personnel and provide technical personnel to establish the operating facility and oversee flight activities.

Included in the TRA/Aermacchi contract were provisions for everything necessary to set up a complete operation integrated into the already existing Sardinian range complex. The overall program was managed out of San Diego by Bob Perkins in his role as International Program Manager for the Firebee Target Director, Pete Petrofsky. Again heading the Ryan target operations contingent was Mike Savino, experienced in setting up worldwide operations, and backed by several Ryan engineers and technicians from both the San Diego and White Sands facilities.

SITE SURVEY OF THE RANGE was completed in early January 1976 and training of the Italian team members was conducted over the spring of 1976 at both TRA's San Diego plant and White Sands Missile Range. Site activation commenced in October of that year and the first of many successful flights was made in March 1977.

According to Bill Berry, Chief of TRA Target Operations, NATO officials were happy with the results of the first Firebee mission ever flown in Europe. The flight was conducted to give the range the opportunity to calibrate and register its tracking and recording equipment.

Three presentations were flown by the Firebee: the first at regular "G" loads and bank angle turns and the next two with execution of "six G" level turns, simulating evasive maneuvers required for the actual firing program. Land recovery by parachute was made in an extremely small area near the launch site.

Mike remembers the land recoveries as "a hairy situation as level ground on Sardinia was hard to come by, and the postage-stamp size area designated for recovery required a helluva lot of skill from the remote control operator. Fortunately we didn't plan to use the recovery site too much, since the Hawk firings were all live. We'd put up 15 birds, and the Hawks would be expected to make 15 kills."

ROUGH TERRAIN OF THE TINY RECOVERY AREA was not always kind *to the descending Firebees. Note the NATO insignia on the vertical fin.*

PREPARING FOR THE FIRST FLIGHT of a Firebee in Europe, *weight and balance is performed in the open.*

WITH WINGS CLIPPED to barely clear the bridge *en route to the launch site, a recently acquired truck replaced the oxen initially scheduled for the towing task.*

FINAL RANGE INSTRUMENTATION proved to be excellent. With four radar stations located strategically up the coast for nearly 100 miles, and with the remote control transmitter up in the mountains to the northwest sending clear information to the remote control station through microwave transmission facilities, this hi-tech installation was in sharp contrast to the centuries-old way of life for the Sardinian natives.

The bridge across the river from the maintenance area to the ground launch site was so narrow that the Firebee's wing tips had to be removed before crossing. And it wasn't until preparations for actual operations were well under way that a truck could be found to move the targets, rather than the oxen for which Mike had been negotiating.

WHILE ALL THREE U.S. military services maintained constant surveillance of Firebee operations around the world, it was the U.S. Navy which recognized the TRA/Aermacchi Salto di Quirra operation as a potential weapons evaluation site for the Sixth Fleet operating in the Mediterranean Sea. Accordingly, a contract went to the TRA/Aermacchi team to supply services for the Navy's surface-to-air missile cruisers, as well as an aircraft carrier. Nine Navy ships, designated Task Group 60.2, were involved in these exercises in mid-January 1978.

Firebees for the Navy operations were flown from the States to Sicily in C-130s, then transported by boat to the Navy support facility at La Maddalena island off the north coast of Sardinia. From there they went overland by truck to the operations site on the southeast coast.

While it was a circuitous route, the journey was simple compared to the chore of clearing those inexplicable foreign no-pilot airplanes through Italian and Sardinian customs.

The first Firebee kill was scored by the USS 'Albany,' flagship of the Sixth Fleet, under VAdm. Harry D. Train, II. The target was flown to the extreme northern limit of the Salto di Quirra Range, approximately 105 miles from the launch site at San Lorenzo, and on the return leg was intercepted and killed by a Talos missile.

Navy use of the services continued periodically until the Hawk firings were discontinued in the early 1980s. With inventories filled and all missiles in inventory tested, the requirement for further Sardinia Hawk missile firings disappeared. The limited use by the U.S. Navy was insufficient to maintain the Firebee capability at Salto di Quirra and Firebee support in the eastern Mediterranean was phased out.

VETERAN BASE MANAGER MIKE SAVINO surveys a survivor *of a Navy Talos missile mission.*

DC-130 LAUNCH PLANE, WITH FOUR WING-MOUNTED FIREBEE TARGETS, banks steeply as it heads out from Point Mugu to the Pacific Missile Range for Navy exercise.

AIRBORNE AND READY TO GO. Nestled under the wing *of a DC-130, Firebees are prepared after launch to act the role of 'enemy.'*

Dave Gossett

PILOTLESS TARGET AIRCRAFT

PTAs at Work

ONCE NEW PTAs *(pilotless target aircraft, to use the new terminology) were being delivered off the production line to user commands in the Army, Navy and Air Force, the task of helping integrate them into routine operation usually fell to the manufacturer's field service experts.*

Later, as the armed services took over complete operational responsibility, the military, as new requirements arose, often contracted with private aerospace companies' service organizations to provide target services staffed with their civilian technicians.

From the continental U.S., operating locations expanded around the world — to the Philippines, to Sardinia, to Panama, to Hawaii, to Puerto Rico, to Okinawa, Korea, Taiwan and Japan.

Even as all eyes were on the war in Vietnam, where secret 'spy' reconnaissance missions were being flown by Ryan **Model 147** *birds, routine target training with standard* **BQM-34A** *vehicles was going on elsewhere to validate the combat readiness of U.S. and allied defense forces.*

Firebee Is — and similar PTAs — originally had, and still have a dual purpose. They serve as realistic targets for **training personnel** *in intercepting 'enemy' aircraft and missiles; and they make possible the* **evaluation of new weapon systems** *designed to destroy hostile air vehicles.*

Originally, drones were launched in mid-air after being carried aloft by mother planes.

Later they were also launched, with the aid of Jet Assisted Take-Off (JATO) rockets, from rails, followed by zero-length installations at hard sites. Next came truck-mounted mobile rail launchers. Navy ships, as well as cargo transport aircraft, also became launching platforms.

Birds were recovered by parachutes which lowered them to the ground or ocean for reuse. Later came mid-air retrieval by helicopters and the use of inflatable bags to cushion ground landings.

Starting out as a rather simple machine in the mid-50s, the jet-powered target drone evolved over the years into a high-tech vehicle as defense contractors improvised new control systems to meet the changing requirements of the armed services.

"Give us your requirements," said the industry's engineers and service technicians, "and we'll figure out a way to do the job." And they usually succeeded.

Thus, over the years, pilotless target aircraft have been adapted for use in an ever-expanding spectrum of sophisticated missions.

Many of the operational techniques and equipments developed for target missions in the '60s, '70s, and '80s led to present-day applications for reconnaissance use and weapons delivery.

What follows is a sampling of accounts – not necessarily in chronological order – of aerial targets filling their vital defense roles.

IN THE AREA OF TRAINING personnel, two systems of defense are involved: (1) air-to-air missile firings during simulated combat conditions, and (2) firings with ground-based or ship-based anti-aircraft weaponry. Both involve defending against attacking enemy aircraft and missiles.

The military services have called attention to their training programs through emphasis given by the Navy to its 'Top Gun' school and by the Air Force with its 'William Tell' biennial weapons competitions.

The ultimate test of a pilot's ability to survive in combat is to escape the missiles, shells or bullets being fired at him, and in turn, direct his firepower at his opponent. Preparation for the actual life-or-death showdown has been years in developing.

The open cockpit aces and heroes of the first World War had no pre-combat training and developed their skills by getting on the tail of their opponent and cutting loose with their machine guns.

World War II didn't do much for enhancing the training of fighter pilots prior to combat. Gunnery ranges were located in remote areas of the California and Arizona deserts where fighter pilots could shoot at sleeves towed by a slow-moving tow plane.

The surviving aces of World War II had much in common with their counterparts of 1918 in that their pre-combat aerial training was minimal. It wasn't until the advent of jet-powered unmanned aerial targets in the early 1950s that pilots could begin to receive training that somewhat simulated actual combat conditions.

In the ensuing years, both the Air Force and the Navy perfected the use of pilotless target aircraft at varying training locations. The Air Force weapons meets at Tyndall AFB in Florida were started in 1958. Named after the legendary Swiss hero who was forced to shoot an apple off his son's head with bow and arrow, the William Tell meetings began a series of competitions involving pilots of the leading U.S. fighter squadrons, both domestic and foreign-based.

Evaluation of surface-to-air weapons, and training personnel in their use, was especially important to the Army. It was also of concern to the Navy in the case of ship-based anti-aircraft installations. Much of the ground-to-air action took place at the White Sands Missile Range (WSMR) in New Mexico.

TOP GUN

Ed Hayes

ONE OF THE MOST spectacular movies of 1986 was 'Top Gun' featuring the incredible technology and training employed by the Navy in schooling its fighter pilots for combat. Campus for the Navy Fighter Weapons School in the early '60s was a metal trailer on the mesa adjacent to the Miramar Naval Air Station at San Diego.

SCRAMBLE! The scramble is on to intercept and 'kill' Firebee I and II Pilotless Target Airdraft.

Teledyne Ryan Aeronautical

In 1969 the 'faculty' moved into a six-room school, and by the fall of 1970 had graduated 93 fighter pilots and radar interceptor officers from the 102-hour course. Included in the course was 27 hours of training in bombing, air-to-air gunnery and air combat flying.

During the early days of the school, Firebee I targets were the simulated enemy and now, several decades later, are still the most-used subscale target in its inventory.

Combat returnees who have faced the realism of survival under enemy attack agree that their combat training paid handsome dividends. Surviving some of the most violent air-to-air fighting in World War II, Pappy Boyington, top Marine ace in the South Pacific remarked, "The air battle is not necessarily won at the time of the battle. The winner may have been determined by the amount of time, energy, thought and training an individual has previously accomplished in an effort to increase his ability as a fighter pilot."

ONE OUTSTANDING contemporary example of the value of air-to-air training comes from Randall (Duke) Cunningham concerning his training against the Navy **BQM-34A** Firebee targets prior to his assignment to the Vietnam theater in 1969-70. During his two tours, he flew 300 combat missions and was awarded almost all of the Navy medals available, emerging from that conflict as the Navy's first all-missile ace.

CDR. DUKE CUNNINGHAM, All-Missile Ace, in 1990 elections defeated the four-term incumbent to become the newest U.S. Congressman representing San Diego.

Recounting some of his experiences with Cunningham, TRA's Ernie Mares, who at one time was a Fighter Squadron 96 pilot serving with Cunningham, recalled: "He's the only man I knew who used 'lag pursuit' tactics to maneuver into position to score Firebee kills with a 'Sidewinder' and then trigger a 'Sparrow' for the coup-de-grace."

In a recent visit, Cunningham said, "The TRA Firebee aerial target systems are the most realistic simulation of threat sources available in air combat training. My F-4 was armed with Sidewinder and Sparrow missiles at that time, just as it was in real combat."

"Realism" is the key to Cunningham's comments. He emphasizes, "Exercising against a realistic threat source simulator like the Firebee is a real asset in preparing for combat."

And what is the bottom line? Navy's Top Gun school says it best; YOU FIGHT LIKE YOU TRAIN!

MILITARY BASES THROUGHOUT the world were the proving grounds for those innovations required by the defense planners for varying target requirements.

One of the major advances in answering the challenge of a more realistic enemy simulation was the development of the Increased Maneuverability Kit (IMK). It was Pacific Air Force fighter aircraft operating out of Wallace Air Station in the Philippines which first flew against the IMK.

Operated in the 'normal' configuration, the Firebee I could make 45 degree bank angle turns, pulling up to 2 or 3Gs. With the addition of IMK, it should reach up to 6Gs and bank angles up to 78 degrees. In terms of realistic performance, the IMK bird was a major departure from a 'canned' mission to one in which the target flew no set patterns.

Random and unanticipated high G maneuvers executed by the Firebee remote control operator required pilots to apply their aerial combat skills to acquire, lock on the target and fire their weapons to score kills on the bird. Air Force pilots, flying F-4 'Phantom' jets were hard pressed to tangle successfully with the new system.

Technical advancements for one service were quickly acquired by other branches of the military. It was at the Navy's Atlantic Fleet Weapons Range (AFWR) that the first quadruple Firebee flight was staged to simulate a mass enemy air attack against units of the U.S. Fleet.

ON LAUNCH PAD IN THE PHILIPPINES, Firebee will soon be airborne to act as a 'clay pigeon' for Pacific Air Force F-4 Phantom jets.

U.S. Air Force

SIMULTANEOUS FLIGHT OF FOUR FIRE-BEES launched from DP-2E aircraft *provided elusive, challenging targets simulating enemy air attack on Navy task force.*

Robert Watts

With two birds equipped with the IMK launched from each of two DP-2E aircraft, the expanded formation of 'enemy' intruders made presentations at different altitudes between 1,000 and 15,000 feet. All four birds were launched within 10 minutes, providing two dual presentations with all four targets airborne at the same time. The missile ships of the U.S. Second Fleet had a good workout!

Vice Admiral B.J. Semmes, Jr., Second Fleet Commander, noted in a message to TRA: "All four targets were programmed for different profiles and each performed flawlessly, adding much realism to the exercise."

It was in the same general area of the Pacific where the Air Force first flew against the IMK Firebees that the Navy also put on a good show. A couple of fighter-interceptor crews (graduates from the Top Gun school) attached to Miramar's Fighter Squadron 92, were awarded kill plaques for their marksmanship. One team made its kill at 15,000 feet altitude after taking off from NAS Cubi Point in the Philippines. Launching two Sparrow missiles from a head-on position, it scored 90 degree hits to destroy the Firebee.

During the same deployment, a second team squirted a Sidewinder up the tail pipe of a BQM-34A at 18,000 feet after taking off from the USS America (CVA-66). The attack carrier was operating off the coast of Okinawa.

WHILE THE IMPROVED maneuverability kit was an advancement over the performance of the 'standard' Firebee, it still had some deficiencies. Time to stabilize after rolling into 5G turns was running around 12 seconds, and G forces and altitude control varied. Both the Navy and the Air Force

were looking for a target that could simulate a threat aircraft capable of rolling into tight turns in a matter of seconds and holding consistent 6Gs while in the turns.

An engineering effort, headed up by veteran Carroll Berner, responded with a system dubbed MASTACS (Maneuverability Augmentation System for Tactical Air Combat Simulation). Several Navy-furnished BQM-34As were modified to accept the new control device, and the test range at Pt. Mugu in Southern California was made available.

Nine MASTACS proving flights were flown on the Pacific Missile Range between January and April, 1971. Again, the Navy's Fighter Weapons School at the Miramar NAS in San Diego was greatly interested in the potential of the elusive bird. Consequently, a graduation exercise was scheduled for May 10 at Pt. Mugu.

Cdr. John C. Smith, commanding officer of the Top Gun school, elected to ride as radar operator and chief tactician as he and three other combat veterans from Vietnam scrambled in F-4 Phantom fighters from Miramar. Both planes were equipped with mixed loads of Sidewinder infrared and Sparrow radar-guided missiles.

What developed was a no-holds-barred contest. Cdr. John Pitzen, Top Gun combat instructor, was tactical director for the Firebee, and instructed TRA's Al Donaldson, who manned the remote control station. In effect, they were in the 'cockpit' of the target, and after the stage was set for a head-on approach, the Firebee proved to be an extremely elusive aggressor.

Open-circuit radio chatter told of the manned aircraft difficulties. Smith called "Tally-ho, off the left wing" but the drone was able to pull such a high-G turn that the F-4 could not follow the maneuver. Smith was learning the hard way that Donaldson and Pitzen could rack the Firebee into a hundred degree bank and make a 180 degree reversal turn in only 12 seconds, permitting the drone to get in behind the now vulnerable F-4. In this position, the drone ceased being a target, but an attack aircraft.

The flight was a convincing demonstration of both offensive and defensive maneuvering by the drone. By going a step beyond the requirements for just a training target, Panadora's box was opened for a quick peek at a potential all-robot Air Force. No comment was made at Mugu concerning the hypothetical case: If the Firebee had been armed with its own missiles, could Pitzen have shot down the Phantoms? Both the Navy and TRA quietly backed away from pushing the concept any further at that time.

Later MASTACS was taken to Tyndall Air Force Base in Florida, but in support of high maneuverability training, rather than an air-to-air combat atmosphere. Still, the learning potential for future robot warfare was apparent.

RUGGED NAVY BQM-34A training target *had already flown 36 missions for air-to-air missile firings by Navy and Marine Corps pilots.*

flight. The four-drone formation join-up was accomplished shortly after launch.

A fly-by formation presentation was then flown (first time around) for air-to-air missile firings. After a good presentation all were recovered at mission completion for refurbishing and further flights.

For the complex mission TRA's Mike Savino was the lead remote control officer assisted by Navy personnel of the Threat Simulation Department.

Top Gun officers and students continued over the years to test their battle-readiness. In April, 1987 they were still shooting at subsonic Firebees, now equipped with electronic jamming devices as well as having formation-flying ability.

In live missile firings, two NAS Miramar based F-14 Tomcat fighters scored Sparrow and Phoenix missile 'kills' just minutes apart. Both missile intercepts were made at distances of more than 20 miles.

TWO NAVY F-4 TOP GUNNERS RETURN FROM ENCOUNTER with Firebees *equipped with maneuverability augmentation system for tactical air combat simulation (MASTACS).*

SHORTLY AFTER THE SHOOTS at Mugu (and later at Rosie Roads) the Navy asked for a drone with a serpentine capability to test evasive maneuvers against their 'Standard' missile. With only minor modifications to the MASTACS, Carroll Berner was able to give them a highly maneuverable target in an Army/Navy cooperative effort at White Sands, New Mexico.

Another quadruple mission was staged by the Navy in 1983, this time at the Pt. Mugu test center. Four **BQM-34S** Firebees configured with J85 engines were air-launched from a DC-130 for a multiple target formation

WILLIAM TELL

THE PRACTICE OF THE MILITARY services spending millions of dollars in training personnel for eventual combat has been the cause for debate among political leaders for almost 75 years. It was President Calvin Coolidge who, in the 1920s, questioned the necessity of the Signal Corps' proposal to buy more planes for their pilots, and suggested "Why not use just one airplane, and let the pilots take turns flying it?".

Has the expenditure for training paid off when the chips are really down? Testimony from combat returnees indicated that the simulated training they received paid dividends in lives saved and contributed to eventual survival if not complete victory.

Following rather inconspicuous beginnings on the desert around Yuma, Arizona, the subsequent Air Force exercises off the Florida coast in the Gulf of Mexico have continued to hone the skills of Air Force squadrons for more than thirty years.

Aside from some early target ranges near Yuma, the Navy has confined its weapons practice areas to missile ranges in the Atlantic at Roosevelt Roads, Puerto Rico; the Pacific Missile Range at Pt. Mugu, and the Naval Ordnance Test Station, both in California; and the Hawaiian area.

Detailed descriptions of each of the William Tell meetings in Florida are available in Air Force archives, but several of them could be classed as 'typical' of the fierce competition and pride displayed by the participating Air Force squadrons.

CHEESECAKE AT TYNDALL. Can this curvacious cupid *really make William Tell?*

BOOSTING THE JET-POWERED 'BIRD' from zero to 200 knots *in seconds, the Jato bottle plume enshrouds the launch pad at Tyndall AFB.*

Eligibility to participate in William Tell competitions is determined on a local basis in the area which squadrons are selected, ranging in geographical locations from the U.S. to Europe, the Philippines and Canada. Having been qualified, locally-based squadrons are air transported to Tyndall – weapons loaders, maintenance crews, radar control units, and pilots, all are on hand – for exercises, varying in length of time, to test their skills.

At the meets, scramble missions were sounded without notice. Radar teams were required to search out, detect, locate, position and direct defending aircraft to intercept points ranging as much as 80 miles out over the Gulf of Mexico. The intruding bad guy in this scenario is the BQM-34A Firebee, and later, the supersonic **BQM-34F** Firebee II.

Heroic behavior in combat is recognized in all services, and awards are given for outstanding courage. The Air Medal, the Navy Cross, the Distinguished Flying Cross and others are proudly displayed by their recipients. However, skillful performance during peace-time generally does not merit as tangible a token of excellence.

Nonetheless, for a time others filled this award void with 'kill plaques' presented to pilots or ships destroying a drone in weapons evaluation meets or in training exercises.

FLYING BULL'S EYE! For years Firebees have challenged the best of the Air Force's sophisticated weaponry.

Ed Hayes

In the '70s, William Tell initiated yet another award. The 'Apple Splitter' trophy was not for individual skill, but for the squadron or unit with the highest overall points in the complicated scoring tally.

The contestants were made up from nine of the top-ranked fighter-interceptor teams in the U.S. and Canada. Rank was not a limiting factor in the competition. The Air Force Chief of Staff, Gen. John D. Ryan, qualified for a kill plaque after making a direct hit with a 'Falcon' air-to-air missile.

THE 1972 SHOOT-OUT at the OK Tyndall Corral was similar to other meetings. However, a new element was added with formal roll-out of the BQM-34F supersonic Firebee II as a competing 'enemy'. The ceremony also commemorated the 25th anniversary of the Air Force.

Later, at Tyndall, five Air Force teams tangled against the subsonic Firebee I, which continued to be the workhorse of the meets. A story published in April '73 indicates one of the reasons for the continued use of the rugged target.

"Tyndall AFB, FLA. -- Like any legend worthy of the name, 'Old Red' refuses to die.

Despite spending 18 storm-tossed hours in the Gulf of Mexico last month, Old Red isn't ready to give up an active service career yet. Instead, the record-setting Firebee target drone is about to launch on its 83rd mission with the Air Defense Weapons Center here.

Delivered in 1964, the plucky Firebee seemed to have 'come a cropper' though, at the end of its 82nd mission when it went down in the Gulf.

Air Force boatmen attempting to recover the target, had to leave it to the mercy of the elements when they were turned back by eight-foot waves and 40-knot winds. However, at first light the next morning a helicopter search team found the persistent Firebee holding its own against wind and weather."

Average flights-per-target at that time at Tyndall were around 20 missions. Following the retrieval from its 82nd flight, Air Force officials finally decided to give the tired old warrior an honorable discharge. 'Old Red' was formally presented to Tyndall's museum for posterity.

While the biennial William Tell '76, held in the first three weeks in November, was a typical meet, it also had a number of 'firsts' recorded during the exercise.

AFTER 82 MISSIONS PLAYING RUSSIAN ROULETTE, 'Old Red' returned *undefeated for a final honorable discharge.*

Dave Gossett

It marked the first time F-4 Phantom aircraft were entered in the competition, the first time a Tactical Air Command squadron was participating, and the first time the Air Force supersonic BQM-34F Firebee II was used to challenge the skill of the TAC teams.

RISING TO THE OCCASION, all of these elements continued the string of firsts during the course of the meet. In their first try, an F-4 fighter squadron based at Elmendorf AFB, Alaska won the coveted Apple Splitter trophy. The squadron registered three kills to take the trophy for the team scoring the highest number of points against the Firebees. The supersonic Firebee II set a new record for the Air Force when three of the birds were successfully launched in one day.

Although there were no kills registered against the BQM-34F supersonic birds, the pilots were extremely enthusiastic about its performance. Actually, conditions were weighted heavily in favor of the Firebee II. The profiles flown were designed for maximum survivability of the target. This included attacks, with the bird flying at Mach 1.2 on a racetrack pattern, using near-miss scoring equipment configured to avoid metal-to-metal contact. Combine that with the fact that these crews were facing a supersonic target for the first time, a clearer perspective of the safe return of the birds is given.

With 22 presentations made by the Air Force BQM-34F, the subsonic Firebee Is also participated, completing 34 flights with 67 presentations for a total of six kills run up by the shooters.

The late '70s saw increased activity for the supersonic target as well as continued use of the subsonic Firebee, but by 1982 the BQM-34A was used only as backup.

TACTICAL AIR COMMAND PILOTS joined other Air Force shooters *at Tyndall to win the coveted Apple Splitter trophy.*

While the augmentation devices aboard the sub-scale targets like Firebees I and II could accurately simulate the enemy aircraft, some training personnel still wanted full-scale droned fighters as targets to give the actual radar signature of the airframe as well as the IR signature of their after-burning engine.

The ever-present hazard of a damaged full-scale droned fighter being returned to the base and shutting down the main Tyndall runways with a crippled bird necessitated construction of an entirely new auxiliary strip removed from the normal active flight patterns.

F-100 aircraft, formerly used to shoot at the Firebees, now became QF-100s, and gave the fighter pilots the most probable reason for using full-scale targets. Putting a Sidewinder up the tail-end of a full sized airplane gave a helluva lot bigger bang and was decidedly more macho than knocking the infrared pods off a Firebee.

SUPERSONIC BQM-34F FIREBEE II is launched on its final flight *for the Air Force at Tyndall AFB, Florida, in October 1989.*

Ryan Aeronautical Company

PREY FOR

THE ARMY'S MISSILES

THE AIR FORCE had William Tell, the Navy, Top Gun. Both featured the effectiveness of their pilots in simulated air-to-air combat with pilotless target aircraft.

However, it was the Army which first provided training in ground-to-air gunnery – the weapon, the 'Hawk' and other anti-aircraft missiles; the target, the **MQM-34D** Firebee.

DURING THE TWO DECADES covering the Korean and Vietnam conflicts, all three services called upon the Firebee to sharpen their skills for actual combat conditions. The White Sands Missile Range (WSMR) in New Mexico was the Army's prime location for the development of the 'Hawk' missile, the 'Chaparral' and other surface-to-air systems.

CARRYING TWO INVENTIVE DECOYS to be deployed in flight, *the Firebee-Towbee combination provides low-cost, expendable targets for heat-seeking missiles, upper right photo.*

Ryan Aeronautical Library

The lethal accuracy of the Hawk made the destruction of the target on each flight impractical so the 'Towbee' concept came into being. Deployed from the wing-tips of the ground-launched target, a light-weight cylindrical body containing a high-intensity infrared source, was released on a cable to be towed behind the mother-drone. The ground-to-air missile would 'home' on the towed target, saving the Firebee for reuse following pin-point recovery on the desert floor.

Design and testing of the Chaparral, the ground-launched version of the Sidewinder, also improved after being fired on hundreds of occasions at the Firebee-Towbee configuration.

Initial launches of the Army's original **XM-21**s at White Sands were accomplished with an 80-foot rail system which was not only expensive, but too frequently unreliable. The next step was the design of a zero-length rail (8-foot) using an 11,000-pound-thrust JATO bottle to get the bird airborne. Mounted in concrete, the rails proved to be a success, but confined launches to a single restricted area.

NECESSITY, they say, is the mother of invention.

Paul Bunner, Ryan's innovative base manager, saw the need for a mobile launcher, but the Army had no funds for the design, engineering and manufacture of such a system. Paul scrounged a surplus Army flat-bed trailer, welded together a set of unused rails, towed his brain-child out on the desert floor and made the first of many mobile launches of a Firebee. Primitive, yes, but the forerunner of mobile launchers much more sophisticated to be used worldwide.

The permanent launch pad at WSMR's McGregor Range was finally constructed and, with the loading crane, 'sugar scoop' air intake regulator mounted, and engine ignitor system all in place, initial launch for the range was not too far off. However, with typical procurement confusion in the early stages of planning, no provisions were made for a direct control station adjacent to the launch pad.

The winter winds sweeping the area could be chilling, and some protection was needed to shelter the direct control operator while getting the bird off the pad. A slight hill directly behind the pad was an ideal spot for the station, and the ingenuity of Bud Miller's crew solved the problem with direct action. Ma Bell will probably never know how she contributed to the defense systems at the missile range, but a midnight-liberated abandoned El Paso phone booth made a satisfactory temporary direct control station.

Billy Sved, then part of the Ryan crew at White Sands, tells of an incident not covered by tech orders. "We had this one bird that recovered prematurely away from the designated area, and left it overnight for chopper pickup the next day. After retrieval, as the target was lowered to the rehab area, a desert fox poked his nose out of the tail cone and then retreated back up the exhaust pipe.

Don Doerr

EARLIEST GROUND LAUNCH OF FIREBEES was from 80-foot rail *at White Sands Missile Range. A later Army Mobile Launcher, below, paved the way for a more sophisticated system.*

Ryan Aeronautical Company

"The darn fox wasn't about to come out. He must have figured anything big enough to move his 'cave' must not be too friendly. We finally persuaded him out with a couple of broom handles, and he took off across the desert like he was jato-powered!"

U.S. ARMY MQM-34 FIREBEE is launched from Pena Beach, Canal Zone, *left. It will be the target for Hawk missile, right, during annual training exercises in Panama.*

Dave Gossett

Dave Gossett

UPGRADING PERFORMANCE

THE U.S. ARMY in the '70s was developing new, more sophisticated surface-to-air missiles. To evaluate their effectiveness, one of them, the 'Stinger,' required a target with speeds in excess of 600 knots. The standard MQM-34D target in Army inventory couldn't hack the speed requirement.

Not that Paul Bunner and his gang at White Sands didn't try several alternatives. The CAE J-69 engine had a temperature limitation of around 700°C., but with some encouragement from superiors, Bunner attempted to hit the 600 knot requirement for a short time on the Stinger's hot leg.

Shoving the throttle to its maximum, the temperature hit over 900°C., but the speed was not appreciably improved. Not only that, but the engine's turbine blades expanded, dragging on the housing, resulting in a scrapped turbine wheel. Obviously, over-revving the J-69 was not the answer.

The second approach to providing the Stinger missile with a 600 knot plus Firebee was to utilize the General Electric J85-4 engine with a sea level thrust of over 3,000 pounds.

The larger engine of the Mod II required placement of the intake in the nose of the bird instead of the normal 'chin' entry beneath the nose section.

Flight tests of the modified bird with the Dash 4 engine were not successful. Direction-

TO CHALLENGE 'STINGER' MISSILE, Army and TRA *developed MQM-34D, Mod II, powered by surplus GE J85 jet engine which added new dimension to Firebee performance.*

al stability problems were evident at speeds of over Mach 0.9. Despite continued redesign efforts, the riddle of inconsistent stability remained unsolved. Because of limited funding and a slipping schedule, the program was terminated.

The Army Missile Command (MICOM) at Huntsville, ever alert to any means of reducing procurement costs, learned of a surplus quantity of GE J85-7 engines in the Air Force inventory. A number of these engines, originally scheduled for installation in the Air Force 'Quail' air-to-surface missile, had been procured. Launched from a B-52 and expendable, the Quail's engines were available as surplus following the cancellation of the Quail concept when improved missiles were procured.

'BOTTLE-NOSE' MQM-34D, MOD II is made ready for Army's 2000th *Firebee flight at White Sands Missile Range. Crew headed by TRA's Paul Bunner, kneeling right, had reason to be proud.*

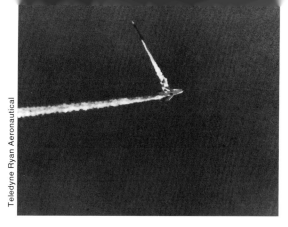

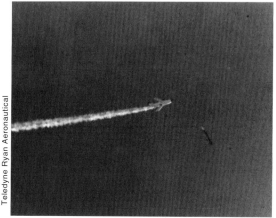

After some negotiation, MICOM obtained four of the GE engines. Several characteristics of the Dash 7 engine made it attractive. In size, it was very close to the J-69, but in performance it was considerably better. With 2450 pounds of thrust, it increased the bird's available power by 44 percent and was calculated to push the Firebee into the desired 600 knot range.

During the turn-around phase of the standard Firebee target, Paul took careful measurements on the standard engine nacelle and of the J-85, and came to the conclusion that installation of the GE engine would not be a major task. Back at the main plant, Larry Emison, using Paul's figures, determined that with a minimum modification to the nacelle the installation was possible. With sufficient hardware manufactured at the San Diego site and shipped to WSMR to accomplish the modification, Paul was able to have a **Model 251** target ready for flight test in early 1973.

Although still designated by MICOM as the **MQM-34D, Mod II**, this model Firebee was also ballyhooed as the 'Supercharged' target, and by May the WSMR crew under Bunner had completed half the required flights for the feasibility tests called for under a new contract.

Results of the initial flights, according to C.D. (Bud) Miller, then TRA program manager on the project, were "Outstanding! Subsonic speeds in excess of 600 knots were attained, and high-G maneuvering turns add new dimensions of 'agility' to the system's performance. This modification offers a broad spectrum of new applications."

The goal of providing a 600 knot Firebee had at last been realized by using the Dash 7 General Electric engine.

The improved performance of the Army's MQM-34D also caught the attention of the Air Force and the Navy, resulting in substantial quantities of targets built to this configuration.

During the first week of the war in the Persian Gulf, America's new 'Patriot' anti-missile missile first saw action, intercepting and destroying 'Scud' missiles launched against Israel and Saudi Arabia by Iraq.

Twelve years earlier, tests at White Sands were used to evaluate the new weapon's ability to locate and destroy Firebees 'standing in' as simulated 'enemy' incoming missiles.

WHILE THE STANDARD BQM-34A had been the Navy's workhorse for many years, 1976 was the year of change. At that time the Navy contracted with TRA to install the

A DECADE BEFORE DESERT STORM, U.S. Patriot ground-to-air missile *proved itself against Firebee target standing in for Russian 'Scud' later used by Iraq.*

newly developed Integrated Track and Control System (ITCS).

Some years later a number of Navy birds were modified to house the GE J-85 engines. With other targets they were air launched from a DC-130 aircraft for multiple target flights on June 9, 1983 at the Pacific Missile Test Center.

Launch of the Firebees was made on schedule, with join-up accomplished shortly after launch. Long range formation presentation was then made for air-to-air missile firings — the first time around. All four Firebees flew well, were recovered at mission completion and returned to PMTC for refurbishment and further flights.

Although performance with the J-85 engine was satisfactory, the Navy later elected to go with the uprated Teledyne CAE J69-T41A. With the installation of the improved ITCS flight control system and the upgraded engine, the Navy standardized the designation of the venerable Firebee model to **BQM-34S**. All Navy production from 1986 to the present continues with the 'S' for all scheduled target shoots.

THE U.S. AIR FORCE got on the improvement band wagon in 1979. Aware of the Army's success with the GE J85-7 uprated engine, they asked TRA to investigate the feasability of using the same engine, but modifying the configuration by rotating the accessory gearcase 180 degrees to the bottom of the engine and replumbing the oil system.

In addition to the engine change, the Air Force joined the Army in purchasing the Microprocessor Flight Control System but retained the old BQM-34A designation for the bird.

Walt Hamilton, TRA Firebee program manager, was satisfied with the results of the Weapon System Evaluation Program (WSEP) concerning operations with the new configuration.

The modified 'A' answered the Air Force requirements for threat simulations matched to the F-15 and F-16 superiority fighters. With improved versions of the Sidewinder heat-seaking and radar-guided Sparrow missiles creating more intense levels in the air-to-air combat environment, the modified Firebees were delivered in February of '89, and successful flights were flown at Tyndall AFB in April the following year.

Survivability of the new bird increased from 5 to 12 flights per target. As Hamilton remarked, "This ability to survive the intense hostile air-to-air combat environment, returning time and time again for target presentations has created cost effective appeal for the Firebee. The system's ability to withstand in-flight missile damages despite its evasive maneuvering capabilities and still fly home to a normal parachute recovery is an inherent design feature."

Fifty of the improved Firebees were delivered by late 1989 and are still the prime threat simulation for the Air Force base at Tyndall, as well as units based in the Pacific and Southeast Asia theater which deploy to the Philippines from Clark Air Force Base for weapons evaluation operation at Wallace Air Station.

NAVY USE WORLDWIDE

THE U.S. NAVY for over forty years has been a major user of unmanned aircraft for training and missile evaluation. Because of its worldwide theaters of operation in widely varying maritime environments, Navy demands on the UAV industry have been stringent.

Dave Gossett

THE VENERABLE DC-130 LAUNCH PLANE with four flight-ready Firebees *beneath its wings leaves Pt. Mugu for yet another Navy missile shoot-out.*

The key operational center for all Navy unmanned aerial targets is the Pacific Missile Test Center at Point Mugu, located on the beach fifty miles north of Los Angeles. With more than four decades of experience in operating target drones, the Target Directorate is responsible for the test and evaluation of all Navy drone systems. Air-launch, ground-launch and sea-launch developments have been pioneered by Point Mugu assisted by contractor technical experts.

Of equal importance to its R&D efforts, the support role of the Center is a vital link in supplying the fleet with enemy threat simulation targets at sea ranges wherever global demands arise. As a staging area for exercises on the Pacific Missile Range Facility, Hawaii, or for fleet-ready operations at the Atlantic Fleet Weapons Training Facility (AFWTF) operating out of Roosevelt Roads in Puerto Rico, Mugu routinely meets the challenge of being on-station and on-time with the required targets.

EXAMPLES OF CRITICAL SUPPORT were the recent Pacific RIMPAC operations and the Mobile Sea Range drills out of Puerto Rico where the defense readiness exercises draw heavily on Firebee targets. Air launch from DC-130s opened the ocean ranges 500 miles from any shipping lanes to provide a realistic scenario for testing the Navy's defense capabilities.

Attaining this capability did not occur overnight. Over twenty years ago, it was the Navy which initiated refinements of the air-launch concept for its various target systems. Originally, the launch aircraft used were the aging Lockheed DP-2E 'Neptunes,' configured to carry two Firebees with two direct control stations aboard.

Since the Missile Center at Point Mugu had relied primarily on ground launch for their targets, Composite Squadron 3 (VC-3), operating out of the North Island Naval Air Station in San Diego, was called upon for any requirements for launches conducted well out at sea. For years the motto of VC-3 was 'Skeet for the Fleet' and their ability to reliably furnish the aerial clay pigeons for the various Navy missile systems at Point Mugu and elsewhere was outstanding.

Early on, the Navy contracted with Lockheed Aircraft Service to produce two airborne control platforms, converting the C-130A 'Hercules' four-engine cargo planes to the DC-130 target launcher configuration. VC-3 inherited the first two airplanes and continued supplying air-launched birds until the composite squadrons were discontinued in 1980. Since then civilian contractor services have operated the launch planes.

SUPPLEMENTING AIR LAUNCH CAPABILITY of Navy at Roosevelt Roads, *Dick Manceau's crew installed the first Navy ground launcher.*

Teledyne Ryan Aeronautical

Ryan Aeronautical Library

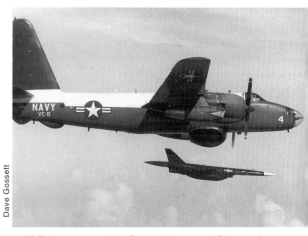

DIRECT CONTROL OPERATOR (DCO) aboard DP-2E Neptune launch plane *runs airborne checks then calls 'bird away' as supersonic Firebee II, below, is on its way.*

Dave Gossett

VC-3 was not alone in providing air-launched target services to the fleet. VC-1 was established at Barber's Point (Barking Sands) in Hawaii in 1966, with Ryan's Billy Sved conducting the initial on-the-job training there. With the development of the RIMPAC exercises in the Pacific, the Hawaiian facility played an important role in fleet target support.

Next to VC-3, VC-8 at the Navy's Roosevelt Roads facility in Puerto Rico was the busiest composite squadron. It was VC-8 which co-developed (with Point Mugu) the first ship-launched drone capability, eventually using a remotely controlled air-sea rescue boat to launch a remotely controlled Firebee.

And, in February 1970, it was VC-8 which worked with Dick Manceau's TRA crew to make the first Navy ground launch on the Atlantic Fleet Weapons Range.

Admiral Semmes commented, "Particularly noteworthy is the ingenuity and forehandedness of the Atlantic Fleet Range Support Facility in the fabrication and initial employment of a BQM-34A ground launch platform that filled the gap created by a critical shortage of drone-launch aircraft." To Manceau's civilian group the Admiral said, "Your extraordinary reliability in meeting operational requirements and the record number of targets provided with a minimum loss ratio is most noteworthy. Well done!"

Sixteen years later, with Libya posing a serious threat in the Mediterranean, it was training at the Navy's Mobile Sea Range, operating out of Puerto Rico, that enabled the Aegis cruisers, the USS 'America' and other U.S. Sixth Fleet battle groups to steam through the Straits of Gibraltar into the Mediterranean and negate the Libyan threat in that area.

During training operations, Firebees had been launched in formation to fly at altitudes from ten feet above the waves to 10,000 feet in order to simulate enemy aircraft attacking U.S. cruisers. Such training proved to be most effective.

Then, in 1990, with the tremendous Naval force deployed to the Persian Gulf and the Red Sea during the Iraqi crisis, the at-sea training provided with Mugu-supplied Firebees enabled our defense forces to be ready to shoot with accuracy.

U NDER NAVY SPONSORSHIP the 'elkhorn' flare system was developed to prevent the Sidewinder missile from homing on the tailpipe of the target. Mounted on the wingtip of the Firebee, each pronged flare holder held three flares which could be fired individually, allowing six passes by the fighter planes to get off a missile without destroying the Firebee.

Teledyne Ryan Aeronautical

WING-TIP FLARES to attract heat-seeking missiles *were circumvented by zealous Marines until the brass caught on.*

It was a Marine reserve contingent from Virginia which discovered that the system could be circumvented. Once locked on, the pilot would give the 'Tally Ho' cry and the flare would be ignited and burn for approximately two minutes. Let's face it. Shooting off a flare was not nearly as spectacular as putting the missile up the drone's tailpipe and watching the grand explosion as the bird disappeared in a flaming cloud of debris. By delaying his firing until the flare was burned out, the Marine reserve pilot could pickle off his weapon and enjoy the show.

Budget-minded Navy procurement soon put an end to these fireworks. If the fighter pilot didn't get his shot off within the two-minute time the flare was burning, he was penalized with a miss and downgraded on his ability. Subsequently, the number of flights-per-bird increased considerably.

Fleet readiness of missile-equipped cruisers was verified at Roosevelt Roads, Puerto Rico, before receiving full approval for combat duty. However, green crews on initial

Ed Precourt

SLEEK, QUICK AND ELUSIVE, a covey of Supersonic Firebee IIs *is made ready for operations based at Roosevelt Roads, Puerto Rico.*

TO MEET THE CUSTOMER'S so-phisticated shopping list *Firebee targets can be equipped to simulate a wide variety of conditions necessary to provide realistic targets for training and for evaluation of weapon systems.*

deployment were not all fully qualified until having the opportunity to shoot a few ship-to-air missiles at the Firebee.

On one occasion, Bud Miller, Base Manager for the Ryan group at Rosie Roads, made four presentations to a newly arrived cruiser. On four passes, no missile was fired, and Bud ordered the bird to head for the recovery area, when he heard the cry from the cruiser "Missile Away!" Due to the fact that the target was already in the recovery chute, it was a no-hit exercise.

The Firebee landed close enough to the cruiser for it to attempt retrieval. Using a one-inch hauser, they hauled the bird aboard, nearly separating its tail feathers. Next step was to remove the battery and carefully wrap it in the parachute with the expected ruin of the chute. However, the ship crew's efforts were not in vain. The skipper was awarded the valued 'kill' plaque for his first exercise.

PRIOR TO THE INTRODUCTION of the BQM-34A targets into the Navy inventory at Roosevelt Roads, surplus F9F fighter planes were droned for air-to-air target exercises. Takeoff, controlled from a direct control booth, was accomplished by heading the plane down the runway. After drone takeoff, control was taken over by two airborne piloted planes, which would pass control of the full-scale fighter drone back and forth in the firing area.

Mission completed, one fighter jock would herd the F9 back to the base, line up with the runway and turn control over to the 'Fox-cart' operator at the end of the runway. Manning the Fox-cart was a somewhat hazardous occupation, since damage to the drone occasionally caused it to veer off the runway and careen into the first immovable obstacle.

Initial Firebee deployment was accomplished similarly, with the target being ground launched; then controlled in the air by the fighter pilots. However, with the advent of efficient ground control equipment installed and operated by Ryan controllers, the duties of the pilots were considerably reduced.

This created one of the early feuds between the use of unmanned vs. manned aircraft. Eventually pilots were returned to their designated task of shooting, rather than using manpower and fuel to herd their prey through the sky.

Firebee operations continue at Roosevelt Roads under the civilian technicians, and the improved efficiency of the remote controlled usage resulted in an increased proficiency of the ships and fighter aircraft.

"CAN DO AT POINT MUGU" has long been the spirit of the Targets Directorate. In 1985, under the leadership of Capt. W.E. Eason and John Goolsby, the staff of 250 Navy and civilian technical experts shared the responsibility in all fields of threat simulation, providing elusive targets for the Navy's most sophisticated air-to-air or surface-to-air missiles.

More recently, Commander Bob Williams, as aircraft targets officer, has been responsible for the task of presenting targets to Navy pilots engaged either in operational training or in the development of new missile systems.

To accomplish this job, Williams had an array of targets, both full-scale and sub-scale, with which to work. The full-scale birds are the droned F-4 Phantom and F-86 Sabre fighters. Both Navy pilot trainees and experienced jet jockeys enjoy shooting at these larger birds because of the bigger bang, but the true workhorse for the Directorates is the Firebee **BQM-34S**.

NERVE CENTER AT THE NAVY'S Pacific Missile Test Range, *Point Mugu, control all aspects of both air-launched and ground-launched unmanned aerial vehicles.*

Last of the supersonic BQM-34E Firebee IIs was shot down in January 1990, but the -34S model Firebee Is continue to provide the most realistic target for the air-to-air shoots.

Marshall Laroce, missile target engineering supervisor, acknowledges that under the command of ground controllers, the sub-scales can fly whatever profile is required — straight and level, weaving patterns or very high-G turns. In a recent article he was quoted as saying, "Sub-scales are more than just a barn for them to shoot at; more than just hulks. They can deploy strips of aluminum foil that confound a radar homing system. They can carry electronics pods to deceive or jam missiles and can launch flares on which heat-seeking missiles can home."

After 45 years of TRA's association with the Navy's Point Mugu operations, the combined years of experience in developing sophisticated UAVs to simulate enemy aircraft and missiles have come to full fruition.

BACK INTO PRODUCTION

WARS, NOT PEACE, have a way of validating the usefulness of new technology. Take the cases of vertical take-off fighters and of helicopters, both of which won their acceptance in the proving ground of actual combat — not in war games.

Vertical take-off fighters did well in research programs but there was no proving ground for them until the Falklands War of 1982, where the British Harriers were effectively used.

For years there had appeared to be only a minor role for helicopters in military operations, but the Korean War of the early '50s changed that.

Only when there was a real need for

them in Vietnam did unmanned Lightning Bug reconnaissance planes have a chance to prove their effectiveness.

In the decade following each war, when there is no military requirement for new hardware, further development of proven technology and production are often stifled. So it was in the late 1970s with RPVs, even in their role as aerial targets. The Firebee production line for such targets was pretty well shut down by 1979.

IN THE ELEVEN YEARS ending 1976, procurement by all users averaged 300 new Firebee Is per year, dropping to an average

A DEDICATED LEADER, Hudson B. Drake, TRA President, *is master of ceremonies at 1986 roll-out of new BQM-34S targets, reopening the Firebee production line.*

of 66 per year from 1977-82. In fact, there had been no new target orders at all for the last three years of that period.

By that time the Navy was beginning to see the need to get some new birds into the production pipeline so they would have jet targets available to maintain combat readiness through continued training in firing against realistic aerial targets.

HUDSON B. DRAKE took over the presidency of TRA at a rather critical time — early 1984. The company had been headed by three presidents in the prior eight years and business, particularly new business, was at a rather low ebb.

True, continuing sub-contract orders for the airframe structure of the Army's AH-64A Apache attack helicopters helped maintain manufacturing capability but the desire to be prime contractor producing products of your own design is a very strong motivation for top management of any company.

As in all business situations, timing is all important — important for the customer as well as for the producer of the product.

Drake, a former White House Fellow, Assistant Secretary of Commerce and executive at North American Rockwell, had just come over to the top job at TRA from TRE — Teledyne Ryan Electronics — where he had been successful for 12 years, the last four as president.

One of the first assignments Drake gave himself was to obtain new orders for Firebee targets.

On the Navy side of the equation was Captain John Shulick, Director of Target Systems for the Naval Air Systems Command. During Capt. Shulick's visit to San Diego in early '84 he and Drake discussed the impending shortfall of **BQM-34S** Firebee targets.

In June, Drake followed up by submitting an unsolicited proposal for multi-year procurement which would serve the dual purpose of reducing target unit costs and getting the recently shut-down production line humming again.

Both knew that a shortage of realistic targets would adversely impact fleet readiness and seriously limit the Navy's capability to adequately test its emerging weapons and defense systems. Recently, too, utilization of the remaining Firebees had been increasing because of the demanding requirements resulting in part from Aegis class cruisers being introduced into the fleet.

As Capt. Shulick later explained, "By 1983, many of us in Washington as well as those in our operating forces knew we needed more Firebees. But there was no budget for their procurement.

"WE TOOK ON A CRUSADE to speak out, addressing budget hearings and other influence sources, working with others in the Navy. Hudson Drake and his TRA team picked up the ball in San Diego and in Washington and made our efforts successful."

New contracts for Navy Firebee production were signed in May 1985, and Capt. Shulick was on hand at the TRA factory a year later to accept delivery of the first units under the new multi-year production schedule.

"The timing," Capt. Shulick said, "was perfect. This system assures us that we'll be able to support requirements during defense readiness exercises for five to six years more using your target systems. Congratulations on a job well done!"

Drake not only won praise from the Navy for his work on new Firebee production contracts, but also from G.W. Rutherford, Teledyne group executive who had originally brought him into the Ryan group of Teledyne companies. It was Rutherford who had helped ease out a prior company president to open up the top TRA job, then moving Drake in from a similar responsibility at Teledyne Ryan Electronics. "We moved some people around," Rutherford explained, "to give Hudson clear sailing and put him in as president."

[Hudson Drake's star continued to rise in the industry. In 1988 he was named to a new key post as Senior Vice President of the parent Teledyne, Inc., organization. — Ed.]

CAPT. JOHN SHULICK, Director of Target Systems *for Naval Air Systems Command, joins Drake as new birds leave TRA factory.*

ONE SIZE FITS ALL! Multi-Mission BGM-34C, *carrying 'Big Red' moniker, could fly strike, recce and ECM routes.*

95

SINCE THE DAYS of the caveman, aggressor and defender — over and over again — have developed new equipment and systems to counter each forward step taken by the opposition.

Radar, for instance, was developed in WWII to counter the Nazi air raids over Britain. The list of such catch-up examples is endless.

In this continuing thrust and counter-thrust of technology, the use of modified Firebee target drones for weapons delivery was an idea long on the 'back burner' at Ryan. Over the years many studies were done by both military and industry planners. After evaluation, some resulted in hardware, test and production — others never progressed beyond the paper stage.

Five successful programs of special interest were the **BQM/SSM**, the A, B and C models of the **BGM-34** multi-mission series, and the **AQM-34V** electronic counter-measure vehicles.

BQM/SSM MISSILE

IN A NOT-SO-ROUNDABOUT WAY, Ryan became involved in weapons delivery as the result of the Israel/Egypt Six-Day War of 1967 when a Russian-made Styx missile sank the Israeli destroyer 'Elath' with the loss of 49 lives.

Even before the Elath sinking, officers of the U.S. Sixth Fleet in the Mediterranean had been deeply concerned about their lack of a weapons system to counter the Soviet missile ships. Under the direction of Dr. Robert A. Frosch, Assistant Secretary of Defense for Research and Development, a ship-to-ship missile to be known as the 'Harpoon,' with a range two to three times that of the Styx, was under study. With a time span of more than five years before an operational capability, something had to be done quickly. Enter the Firebee.

For several years, Ryan had been studying and proposing FLASH, its projected Firebee Low Altitude Ship-to-ship Homing missile. Here, the initial research had been accomplished and the quick reaction capabilities of the company were recognized. Unlike solid propellant missiles with a limited range of around 35 miles, FLASH would be a cruise missile with a range capability of more than 100 miles.

Instead of developing a system to destroy Styx in midflight (a concept that came into being nearly fifteen years later with the Aegis capability), the Ryan-proposed Surface-to-Surface Missile system would counter the Russian threat with the ability to deliver a warhead at low altitude and long range, providing an effective counterthreat. The Russians would be out-gunned before they could start to launch.

FOUR CAPABILITIES were essential to convert a standard BQM-34A Firebee target system into a weapons delivery vehicle. Fortunately, both Ryan and the Navy were well prepared for a quick response to a deadly threat. The capabilities included: (1) a demonstrated ability to carry a weapon; (2) terminal low-altitude control to point of contact with the enemy target; (3) ability to launch from a destroyer or other ship under way and (4) real-time guidance to find and destroy that target.

Three years earlier, Ryan had demonstrated to the Army Missile Command in New Mexico the ability to get the Firebee airborne with a thousand pound load of bombs. Using the ASROC rocket booster and adding extended wingtips, that capability was proven.

In response to the second element, the low-altitude capability of the Firebee had been over that route several times. The initial pass at low altitude flight in the early '60s was made with the BLACS, or Barometric Low Altitude Control System. BLACS was a grand

1 - ABILITY TO CARRY A WEAPON. To handle 100-pound overload, larger ASROC booster and extended wing tips provided the capability.

U.S. Air Force

2 - TERMINAL LOW ALTITUDE CONTROL 50 feet above waves *was demonstrated by Firebee equipped with RALACS - radar altimeter low altitude control system.*

effort, but lacked the accuracy required, and the slow response time of the barometric device caused some of the earlier flight test models to plow a rather large hole in the ocean off Pt. Mugu.

RYAN'S RADAR ALTIMETER Low Altitude Control System(RALACS) had been under development since 1965, and became operational on a Firebee by 1966. With a radar altimeter to measure altitude above the ocean and with an electronic output to the control system, the remote controller could make precise and instantaneous changes in the flight path. Expertise in the radar altimeter field was no new project for Ryan. When Neil Armstrong made his soft landing on the moon, his spacecraft was equipped with a Ryan-built radar altimeter.

As to the third demand, the ability to zero-length launch a Firebee from a ship at sea had been demonstrated many times previously. Tests, both at the Navy's Pt. Mugu range and the Atlantic Range at Roosevelt Roads had proven the concept of ship-launch. Later work with the Japanese and their 'Azuma' target launching ship proved that capability beyond a doubt.

Lastly, real time guidance to find and destroy the target necessitated some research and flight testing. The Naval Ordnance Test Station (NOTS), China Lake, California, was

given the chore of developing a 'target seeker' and chose to use their previous experience by installing a TV camera mounted in a BQM-34A Navy Firebee. Several runs over the desert terrain in 1968 successfully demonstrated the concept. The Firebee was made to respond to a proportional control system where the movement of the miniature control stick at the remote control station gave the drone's aerodynamic system the same proportional control inputs.

The controller became a ground-based pilot. Watching the TV, he could accurately fly the Firebee at low altitudes, just as though he was buzzing the desert in a fighter aircraft.

An additional capability had also been demonstrated at NOTS, as a lightweight warhead for the BQM-34A 'missile' was essential.

The prototype of an explosive warhead, using a fuel-air mixture, was under development. It was installed in the nose section of the Firebee. The test bird, mounted on a cart, was then driven by a rocket engine along a rail into a simulated ship compartment with devastating results.

3 - ABILITY TO LAUNCH from a converted aviation rescue boat *confirmed ship-under-way potential operation.*

4 - REAL-TIME GUIDANCE to find and destroy the target *was provided in tests which demonstrated use of nose-mounted TV camera.*

IN SEPTEMBER 1971, with many governmental experts on hand, a pre-contract demonstration of the **BQM/SSM** system **(Model 248)** was staged off southern California.

It was a near perfect test except for the slow response time to bring the bird down from 75 to 35 feet. Before the override command could be given, it flew over the target ship at about 70 feet off the water, striking a wire from the mast which clipped off a wing, ruptured the fuel tank and caused the bird to explode. Had it been armed, it would have rated as a full kill.

The second flight, on September 2, was classroom perfect. Launch at 5,000 feet from the Navy DC-130 was made and the Firebee 'missile' brought down to 220 feet, then 75 feet and on final approach, down to 20 feet right into a perfect hit on the decommissioned destroyer USS 'Butler.' Photo coverage of the impact clearly demonstrated the ability of the SSM to demolish any enemy ship of corresponding size.

Ernie Moser was a TRA 'pilot' aboard the control helicopter, and veteran Al Donaldson was backup controller on the ground at San Clemente Island in the event the chopper lost control of the bird, but the entire mission was handled from the air.

As usual, funding for a program of the size envisioned for the BQM/SSM system was a pacing factor. More significant perhaps was the fact that this system might slow down or curtail development of the more sophisticated 'Harpoon,' or possibly result in its termination. It was apparent that top Navy officials wanted the Harpoon so badly they could not approve an interim program that might jeopardize it.

The vacillating decision was to postpone any final action on either program for another six months. As Bob Schwanhausser so succinctly put it, "The operation was successful, but the patient died."

DEFENSE SUPPRESSION

ALTHOUGH THE BQM/SSM PROJECT had apparently been laid to rest, the demonstrated weapons delivery capability was resurrected as the **BGM-34A** under a different nomenclature and sponsor.

Once again trouble in the Middle East created the need for a response, this time to Russian armament in the control of an Arab nation.

With the Navy's decision to back off from SSM, TRA was in an excellent position to respond to the new Air Force requirement which traded the passive target role of the Firebee for a deadly weapons delivery strike system (Ryan **Model 234**).

Israeli Prime Minister Golda Meir was more than a little upset about the Russian SAM and AAA batteries which the Arabs rushed into place in August 1970 along the west bank of the Suez canal. Seeking help from President Nixon, she learned that our Defense Department was reluctant to provide F-4 Phantom jets to blast the enemy sites because of the high mortality rate to be expected. Added to Golda Meir's situation, there was additional concern from NATO countries in Western Europe that the same threat of multiple, overlapping SAM sites co-located with extensive anti-aircraft artillery installation existed. As a result of these pressures, action by the U.S. was dictated.

The Department of Defense was somewhat leary of sponsoring an air-to-surface unmanned vehicle posing as a missile, and politely referred to the proposed development plan in 1970 as 'Defense Suppression.' Asking for emergency funds of $14 million, six different tasks were undertaken under the code name of 'Have Lemon,' with task 05 going to Teledyne Ryan to come up with four vehicles capable of air-to-surface weapons delivery.

THE GO-AHEAD CONTRACT from Air Force's Systems Command was received in March 1971 and Bill Helmich Jr., a gung-ho TRA engineer, was named program manager. Drawing on the experience and equipments generated over the past several years, Hemlich and his crew submerged themselves into a massive 'Tinker Toy' program. Basic vehicle for conversion to **Model 234** was the **Model 147S**, a survivor of the Vietnam con-

TV CAMERA IN THE NOSE *has zoom lens to transmit image of terrain ahead. On the wing is Maverick electro-optical seeking missile*

Ed Wojciechowski

Teledyne Ryan Aeronautical

'STUBBY HOBO' MISSILE is mounted on left wing of 234 which, in turn, is suspended from DC-130 pylon before launch of the UAV and the weapon it carries.

flict. Several of the four birds already had better than 20 flights over hostile territory and were already equipped with the MCGS and MARS systems.

Helmich recalled, "We combined parts and pieces from six different special purpose aircraft after we'd gutted and refurbished the basic vehicles. We used NA wings with bomb shackles for the hard points to support the missiles. We used the football-shaped antenna system from the TE for the television data link and the MCGS. We used the SD horizontal tips, NC end plates and an S nose with a standard BQM nacelle." Yet, despite this extensive facelift, the classic blood lines of the workhorse Firebee were still easily recognized.

Five captive flights at Edwards AFB prior to October were flown, followed by ten free flights. Two birds were lost during this test period, but by December 14th, the off-site test group was ready for the first full demonstration of the defense suppression capability of a drone firing a powered, guided air-to-surface missile against a simulated SAM site.

All flight work was done by the 6514th Test Squadron, with the TRA team headed up by veteran Gene Juberg. "We actually hit the target," Helmich related, "nine months and ten days after go ahead. When I say 'hit the target,' I'm not using a figure of speech, but referring to the demonstration — and that flight was completed four weeks ahead of schedule."

Juberg described the first flight this way: "In the nose was a TV camera equipped with a zoom lens, which in real time transmitted an image of the terrain ahead to the screen in the remote control van. On the drone wing was an AGM-65 Maverick electro-optical seeking missile also capable of telemetering back to control the actual video of the seeker head as it locked on the target.

"The remote control operator on the ground was able to see the target on his TV screen about five miles out without using the zoom capability. At about 2-1/2 miles out the missile locked on the target and seconds later it was fired under its own power, hitting the target squarely in the center about nine seconds after firing." It was a historic first missile launching from an RPV to score a direct hit. To commemorate the event, the 6514th Squadron was given a distinctive award of commendation.

Seven weeks later, in February 1972, an air-launched BQM-34A itself became a launch platform, this time for the 'Stubby Hobo' missile. As an electro-optical glide bomb with an autopilot which drives the vaned control surfaces to give it guidance, the Hobo hit the target squarely in the center nine seconds after firing.

With proof of concept clearly established, the Air Force took the next step toward establishing an operational concept.

SIMULATING HEART OF A SAM SITE, obsolete radar control van becomes target for Maverick (left photo) and 'Stubby Hobo' (center photo).

Ryan Aeronautical Company

ON THE DAY after Christmas 1971, intense, widespread bombing of North Vietnam (the sharpest escalation of the air war since saturation bombing had been halted three years earlier) was resumed with hundreds of fighter bombers blasting North Vietnam for five straight days. In the ensuing weeks, for the first time in many months, a significant number of U.S. planes and pilots were being shot down.

Such large losses again added new fuel to the need to develop the unmanned defense suppression capability. In the later part of January, TRA was specifically requested by the Air Force to indicate how many birds with a missile carrying capability — primarily the 147SC and SD models — could be rapidly deployed to Southeast Asia.

Bill Helmich recalled, "The philosophy of the Tactical Air Command was to use an RPV attack system to go in on the first wave and soften up the target so that the manned aircraft, F-4 Phantoms and F-105s can go in and finish the job with the human eye. We wouldn't expect to replace the manned aircraft or the pilot. What we really want to do is go in and soften up the targets and give the pilots a fighting chance.

"The drone runs about one-tenth of the cost of modern manned jet fighters which carry one or two pilots each. And, everyone wanted to cut down the number of guests at the Hanoi Hilton, and this was one way to do it."

The excellent camouflage of the SAM sites by the Viet Cong eventually made visual contact — either manned or with the unmanned drone carrying a TV camera — next to impossible. Logically, the next step in improving the strike capability of the Firebee was to install all-weather, day and night IR systems that could spot the SAM installations without visual acquisition.

BGM-34B STRIKE

WITH THE PULLOUT from Southeast Asia completed, it was two years before TAC was able to resume the development of the Firebee strike capability. Following a year's engineering and product improvement effort, the first of eight new pre-production strike birds, designated by the Air Force as the **BGM-34B** (and **Model 234A** by TRA) were ready for presentation to the Air Force.

February 1973, marked the formal roll-out of the new RPVs. Col. Ward Hemenway, Aeronautical Systems Division representative, and other Air Force dignitaries were on hand for the occasion to witness the new bird, complete with scowling countenance and weapons

slung under its stubby wing as the usual acceptance speeches were made.

Initial flight tests for the eight modified birds were flown at Edwards Air Force Base, but shortly thereafter DoD, in the wake of announced closings and reduced operations at bases across the country, relocated the Drone Test Squadron from Edwards AFB to Hill AFB in Utah. Packaging all of the Air Force drones into a single complex, resulted in important savings to the defense budget and opened Utah's Dugway proving grounds to extended areas for test and development of new systems.

[Among the aircraft assigned to the test squadron were Teledyne Ryan's AQM-34 (models L, H, M, T, Q, and R); the BGM-34B and the BQM-34F; YAQM-34A and the FDL-23, a modification of the TRA drone by the Air Force's Flight Dynamics Laboratory to test future RPV maneuvering techniques.]

Already present at Hill when the BGM-34B arrived with its own contractor-operated team under Gene Juberg was Billy Sved, working with the 6514th drone squadron whose combined efforts led to the worldwide motto of "Have drone, will travel."

The BGM-34B differed from the 'A' version in that it had a larger engine, modified tail unit and enlarged control surfaces, giving it greater operational capability.]

Flight missions over the Dugway proving grounds included single and multiple passes against a variety of targets; launching a number of live and inert weapons which included SPASMS (self-propelled air-to-surface missiles) and the Hughes AGM-65 Maverick TV-Guided missiles.

BGM-34B COULD CARRY A VARIETY OF WEAPONS *including SPASM, a self-propelled air-to-surface missile.*

U.S. Air Force

Bud Wolford

POSSIBLY THE WORST CASE of uglies in the history of the many versions of the Firebee was the Pathfinder version of the BGM-34B, said to look like a sad hippopotamus. The six-foot elongated cylindrical nose hanging on the forward edge of the fuselage had three circular openings to house a laser designator and low-level light TV camera in a nose pack designed by Philco-Ford.

Technically successful, its ungainly appearance caused one member of the flight test team to remark "Built by Ford, and it kinda reminds me of the Edsel with that 'lemon-looking' nose, but unlike the Edsel, this thing is doing a good job!"

In November of 1974, a total of six demonstration flights were conducted for the Federal Republic of Germany. The tests were conducted under a U.S./FRG cooperative project called 'Coronet Thor.' Object of the joint R&D activity was to demonstrate feasibility of strike RPVs like the BGM-34B in weather and terrain peculiar to Germany. All missions were conducted in a 30 day period and all test objectives were successfully attained.

"I MAY LOOK LIKE PINNOCHIO or a sad hippopotamus, *but I could be the best Pathfinder in the recce business," so says this 'worst case' BGM-34B.*

BGM-34Bs AT HILL AIR FORCE BASE are shown with an array of ordnance *they are capable of carrying. Pathfinder RPV rear and TV-guided daytime version in foreground. Weapons include Stubby Hobo, Maverick and both MK 81 and 82 munitions.*

Bud Wolford

'FIRE FLY' WAS THE pre-Vietnam program developing the reconnaissance spy plane capability under Big Safari procurement policies.

Still undecided at the time was the 'roles and missions' question of which Air Force command would accept operational responsibility for the recce drones.

Lt. Col. Lloyd Ryan, who had muscled the early quick reaction procurement effort, took upon himself the task of persuading the Air Defense Command to select a sponsor.

Big Safari was not an operating command, yet often handled operations for as much as six months while training the command which would actually use the vehicles it developed.

"After we built the first vehicles and had a very, very limited operational capability," Lloyd Ryan later explained, "we put together a briefing of what we had, and went down to Tactical Air Command Headquarters. We offered the drone program to the Commanding General of TAC. He wanted no part of unmanned aircraft so then we went up to the Strategic Air Command Headquarters at Omaha.

"At Omaha we briefed a few people, but none of them wanted anything to do with it either. Finally we went back to an old friend by the name of Major General William H. (Butch) Blanchard, Director of Operations.

"When we made our presentation to him he said,'Why, of course, SAC would be more than happy to take it – give it to us and we'll run with it.'" That put SAC in the recce drone business for 12 years even though they didn't at first have any DC-130 launch planes in their inventory.

Four years after its turndown of operational responsibility, Tactical Air Command was forced to do a 180-degree turn in their attitude toward unmanned aircraft. Meanwhile SAC, in assuming operational control, had obtained their launch aircraft and was successfully flying the recce birds over enemy positions in Vietnam.

In late 1967, the Air Force asked Ryan to determine if it was feasible for a drone to fly a pre-strike ECM (electronic counter-measures) mission in support of manned aircraft strikes conducted by the Tactical Air Command. Too many aircraft and crews in Vietnam were being lost. Something was needed – some way of dispensing chaff – to jam enemy radars and dilute their defensive capability.

Once again, another crash program was taken on by Ryan. 'Combat Angel' was given a 90-day target. Lt. Col. Red Smith again went to work

SUSPENDED BENEATH DC-130 LAUNCH PLANE, this
147NA *in turn has its own pylons from which two chaff-dispensing pods are hung.*

with the Ryan team. Some 400 odd ALE-2 chaff dispensing pods and a fleet of proven 147NA-configured drones with their launch planes were rounded up. The stubby NA with pods was ready to pave the way for the TAC jocks on bombing missions.

By the time the ECM squadron was qualified and ready to be moved overseas, several months had passed. Then, starting November 1, 1968, President Johnson made one of those decisions that characterized the administration's concept of warfare.

His decision to halt bombing above the DMZ in North Vietnam eliminated the necessity for the droned ECM birds to protect TAC fighters, and obviously gave much aid and comfort to the VC. At that point, the panic effort to build a drone-jamming capability lost its cause. Again, as Red Smith explained, "It did show us a way to respond to a national need to build a set of hardware in 90 days that would fly and do the job."

Thus, the pre-strike ECM capability went quietly into its moth-balled cocoon.

AQM-34H VETERAN OF VIETNAM became test vehicle *at Dugway Proving Ground to test ground launch of bird's ability to carry pods which dispense chaff to confuse enemy radars.*

UP-DATED VERSION OF 'COMBAT ANGLE,' *the AQM-34V demonstrated its ECM capabilities to the Air Force, Army and Marines at 'Gallant Eagle' exercises over the Gulf of Mexico.*

U.S. Air Force

THE COSTLY LESSONS learned by TAC in bombings north of the DMZ in Southeast Asia were not easily forgotten. Chaff dispensing fighter aircraft were extremely vulnerable to the SAM sites, and by the time the unmanned ECM birds were ready for deployment their use was negated by the 'no bombing' edict of the DoD. However, the value of the RPVs in terms of lives saved caused top TAC planners to continue to push for their revitalization.

Under the Air Force's Aeronautical Systems Division, Ryan was awarded a contract in September 1974 to modify veteran AQM-34H and J models returned from Vietnam to a new standardized ECM version to become known as the **AQM-34V** (company Model 255). The modifications improved the electronic warfare, flight control, launch, and recovery systems. This program was dubbed the 'Combat Angel update modification program,' with responsibility given to the 355th Tactical Wing at Davis-Monthan Air Force Base, Arizona.

By April of the following year, a contract was issued for the purchase of 16 AQM-34V Combat Angel follow-on production birds. The 6514th Test Squadron, with their growing record of outstanding operational reliability, was given the initial flight test responsibilities for the new bird and on May 13, 1976, the first free flight was made over the Utah range.

By October, the 6514th had completed the entire test program which demonstrated the "V"s improvements in active and passive ECM jamming, flight control, launch, recovery, and

multiple vehicle control capabilities. Having been satisfied that the new version could handle its assigned missions, operational control of the AQM-34V was turned over to the 11th Tactical Drone Squadron out of Davis-Monthan, with the gung-ho 432nd Drone Generation Squadron doing the key maintenance work.

THE NEW ECM VEHICLE was thoroughly tested by TAC in an exercise named 'Gallant Eagle' conducted at Eglin AFB, Florida during the week of October 30 through November 3, 1978. M. E. (Gene) Juberg, TRA's ops manager for the exercise, summarized the purpose and results of the show in a memorandum to the main plant after completion of the maneuvers. It reads, in part: "This trip report covers the writer's observations of the 432nd TDG participation. Their deployment team consisted of 134 personnel operating out of Hurlburt Field, four miles west of Ft. Walton Beach.

"Gallant Eagle was a combined Air Force, Army, Navy, and Marine exercise. The 432nd's participation was to fly four EW sorties with AQM-34V vehicles on Monday, Wednesday, and Friday.

"The squadron deployed from Davis-Monthan with three DC-130 launch planes with four drones uploaded on each, three CH-3 MARS recovery helicopters, a TPW-2 Ground Director and a minimum of ground-handling equipment and spares. Two additional DC-130 aircraft were flown in from Davis-Monthan with spares for the launch planes, and a third came in at

103

the end of the exercise to transport the Ground Director back home.

"The EW mission for the exercise was to lay a chaff corridor at 3,000 feet MSL while heading toward the coastline. The two drones, flying two minutes apart in trail formation, were then climbed to 19,000 and 20,000 feet MSL respectively, where they took up an active jamming orbit. The drones were followed in by an E6A or EB-57 aircraft which also dispensed chaff and provided EW active jamming through the chaffed corridor."

This operation clearly demonstrated TAC's philosophy of sending in the decoy drones first to protect the manned jammers and fighter planes to follow.

Gene Juberg's field report concluded, stating that all objectives of the meet were met with unqualified success from performance of the AQM-34V. The 432nd Tactical Drone Squadron, under the command of Col. James Witzel, was recognized by top TAC generals and commended for a job well done.

XQM-103

AIR FORCE DYNAMICS LABORATORY conceived and flew *the further-improved ECM bird. XQM-103 (FDL-23) had TV in nose, putting a ground-based pilot in the loop.*

U.S. Air Force

WITH WORK ALREADY under way in 1974 on the AQM-34V tactical electronic warfare RPV, the Air Force Flight Dynamics Laboratory at Wright-Patterson AFB was interested in several potential improvements that could be incorporated in unmanned aircraft.

First consideration was to increase the range of the 'V' birds by installing the larger Vietnam-era 147G wing on the V fuselage. The second area for investigation was putting a pilot into the loop by providing a TV camera in the nose of the bird. Third part of the program was to obtain maneuverability comparable to that of a manned aircraft.

The TV video relay to the remote station would create, in effect, a modified Link Trainer for a pilot to remotely fly the vehicle.

Installation of these systems in the **XQM-103** (FDL-23) research vehicle proved to be no major problem. For research into high maneuvering, structures were strengthened to withstand

10g. However, flight tests indicated additional study was needed on the gyro system for the camera.

It was work accomplished in increasing maneuverability that really paid off. The ability to maintain constant altitude in 6g turns was proven, as was three-axis control identical to that of a manned aircraft.

As program manager for the experiments, Lou Pico, TRA electronic expert, supervised the design and installation of the equipment, and in cooperation with the 432nd TAC Drone Group at Edwards AFB, began a series of flight tests.

To provide data for a variety of future RPV programs, 24 flights were scheduled. However, with the TAC axe falling on all RPV programs in late '75, the 103 was quietly shelved after only a relatively few successful flights.

In spite of the cancellation, the improved flight control box, eventually known as the Microprocessor Flight Control System (MFCS) managed to survive. Aware of the improved capabilities in the system, TRA was able to interest the U.S. Army, and was awarded a contract for its incorporation of MFCS in Army MQM-34D targets. Features of this control system have since been incorporated in almost every TRA UAV.

SAC to TAC

THE 4080TH STRATEGIC Reconnaissance Wing (SRW) moved to its new headquarters at Davis-Monthan AFB at Tucson, Arizona in July 1963. Less than a year later it was deployed, together with its U-2 manned aircraft and Ryan Model 147 unmanned aircraft to Southeast Asia.

There, the Wing, redesignated the 100th SRW in 1966, operated until after the war, when remaining assets and personnel were returned to Davis-Monthan.

Back at Davis-Monthan another Vietnam unit, the 11th Tactical Drone Squadron (TDS), was testing electronic countermeasure RPVs. The fact that two RPV groups – one Strategic, the other Tactical – were at the same base under separate major commands did not go unnoticed.

In July 1976 they were combined under the Tactical Air Command, operating as the newly formed 432nd Tactical Drone Group. Its 11th Tactical Drone Squadron operated the AQM-34V Combat Angel ECM birds. The Vietnam-weary reconnaissance workhorse AQM-34Ms and AQM-34Ls, were operated by the 22nd TDS. [The -34M(L) had Loran navigation gear; the -34M(TV) version had realtime TV capability].

MULTI-MISSION BGM-34C

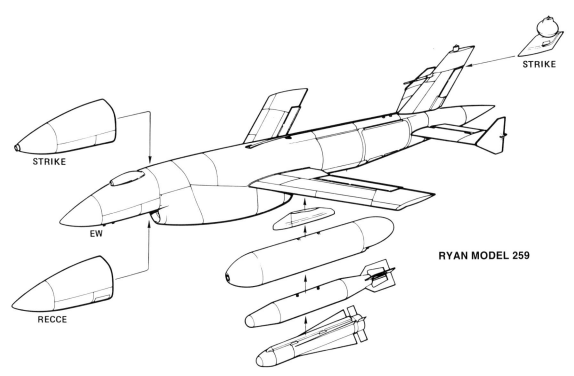

STRIKE

STRIKE

EW

RECCE

RYAN MODEL 259

STRIKE

RECONNAISSANCE

ELECTRONIC WARFARE

THREE DIFFERENT BIRDS; three different missions.

Why not combine all three capabilities into just one vehicle?

Such was the genesis of the **BGM-34C** Multi-Mission RPV, the first of which was rolled out for test in 1976.

To some extent, technical and operational rationale for a single multi-mission vehicle stemmed from a quarter-million dollar study contract the Air Force had awarded TRA in 1972.

The Air Force development philosophy for the BGM-34C was to integrate demonstrated technology from other programs. To accomplish that end a development, integration and flight test contract was placed with TRA as systems integrator in November 1974.

By use of three interchangeable modular noses, the basic capabilities of the AQM-34V electronic warfare, AQM-34M (147SD) recon-

naissance and BGM-34B air-to-ground strike RPV, the latter with TV-guided weaponry, were brought together in one vehicle. The single modular payload RPV was expected to reduce vehicle life cycle and maintenance costs.

Five Ryan model 147SC (AQM-34L) RPVs were modified with structures strengthened for ground as well as air launch. Lear Siegler would provide the updated automatic guidance electronics (AGE) from the **YAQM-34U** version of the 147 RPV series.

Remote control during flight was provided by a specially modified Lockheed DC-130H launch aircraft with Sperry Univac guidance system able to control up to eight RPVs simultaneously.

Ability to ground launch from a mobile launcher was designed into the vehicle. Recovery of the birds would be on land, water or by mid-air retrieval during terminal descent stages.

Because of the prospect for an eventual long-range program, TRA assigned key aerospace engineer Hugh B. Starkey as its Director of Tactical RPV systems. Col. Ward W. Hemenway, Air Force Deputy for RPVs, supervised the program for the Aeronautical Systems Division.

USAF BGM-34C MULTI-MISSION RPV

Robert Stewart

MULTI-MISSION RPVs *"should reduce the loss of fighter aircraft....and the lives of members of the crew." Note duct for TV camera aft of nose cone, and ECM pods under wing.*

WHEN THE FIRST of the BGM-34C RPVs was rolled out of TRA's San Diego factory in August '76 it revealed some external changes from previous strike vehicles, including an open duct for the TV camera installed aft of the nose cone.

At the roll-out ceremony TRA President Barry J. Shillito noted that the multimission vehicle was a "significant step forward in our effort to maintain an adequate defense posture at an affordable cost.

"An even more important economy comes from the opportunity to reduce the losses of fighter aircraft and above anything else, the lives of members of the flight crews.

"In the case of electronic warfare missions, the RPVs will fly ahead of the manned strike aircraft, dispensing chaff and electronically jamming surface-to-air missile system radars. By degrading defenses, impressive reduction in manned aircraft attrition can be accomplished."

FIVE OF THE PROTOTYPE vehicles were delivered to the Hill-Wendover-Dugway test range adjacent to Hill AFB in Utah for an extensive flight test evaluation. Future plans called for procurement of some 20 production vehicles per year starting in July 1977.

Al LeBaron headed up TRA's 15-man field test crew working with the 6514th Test Squadron which had recently moved its flight test operation from Edward AFB to Hill AFB. Elements of TAC, the user command, would also be involved in the test program with the first five flights conducted by contractor technicians and engineers.

The first flight was made over the Wendover-Dugway test range a month after the San Diego roll-out. Purpose of the initial 44 minute, 300 mile flight was to demonstrate the airworthiness of the vehicle, and to check flight control and other on-board systems. The BGM-34C was in the reconnaissance configuration for the first flight.

The second test flight of BGM-34C was made in October in the electronic warfare configuration. All test objectives were accomplished.

Not unexpectedly, contractor personnel during the five flights they flew occasionally found themselves in the middle between their direct customer, the Aeronautical Systems Division; the 6514th Test Squadron; and the designated user command, TAC, which would later have operational responsibility for the production multi-mission vehicles.

Also, TRA had two technical groups at Hill AFB at that time – LeBaron, with the multimission program, and Billy Swed, with a group providing routine interface between the 6514th and the company on a number of continuing programs.

the wide band video data link subsystem of the strike configuration.

Three birds were sacrificed during test, only one to an in-flight malfunction. The other two resulted from mid-air retrieval accidents.

A highlight of the test program was a November 1977 demonstration of RPV formation flying. Two BGM-34C RPVs in the electronic warfare configuration carrying ALE-38 chaff pods flew in formation for 50 minutes while performing an electronic warfare mission profile which demonstrated that unmanned vehicles could lead a strike force penetration to targets in enemy territory.

Launched at 15-second intervals, the two birds, with modular EW noses installed, navigated independently on their own internal systems while maintaining an approximate 750-foot lateral separation for the duration of the flight. It was an excellent demonstration of the high degree of accuracy attained with the computer based LORAN navigation system.

On completion of the test program an Air Force News Release announced that "the next step in the acquisition process will be a production decision by the Department of Defense." At that time, proposals were being considered for possible procurement of 145 vehicles over the next six years. However, planned production did not follow.

The multi-million dollars 'budgeted' for initial production of BGM-34C systems were diverted to buy tactical aircraft spare parts.

ON THE WING AND READY FOR LAUNCH are AQM-34V **electronic countermeasure RPV,** *nearest camera, and a* *BGM-34C multi-mission strike vehicle.*

THE 18-MONTH FLIGHT TEST program, consisting of 27 flights, was completed in April 1978. Fourteen flights tested the reconnaissance version, 12 the electronic warfare configuration, and one flight evaluated

YEARS AFTER ITS CONCEPT, **and with numerous versions** **built and flown,** *the ultimate* *BGM-34C could provide a UAV* *capable of strike, recce and ECM* *missions. Note variety of nose* *and pod configurations.*

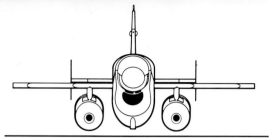

WHAT NOW?

IN SUPPORT OF 'LINEBACKER II,' the B-52 heavy bombing of North Vietnam in January 1973, the 100th SRW flew a heavy schedule of reconnaissance and bomb damage assessment (BDA) missions with AQM-34L and -34M vehicles.

By mid-month came the official cease fire, whereupon the recce drone operation was placed on a 'hold/standby' basis.

Some follow-up validation reconnaissance and communication intelligence (Comint) flights were continued into summer 1975 with AQM-34M (147SD) and AQM-34R (147TF) vehicles.

Meantime, most of the assets from Southeast Asia were returned to the States and warehoused

However, new up-dating projects, such as BGM-34A, B and C strike birds, and the pre-strike AQM-34V 'Victor', were under way to refine state-of-the art technology.

These programs and all reconnaissance vehicles came under Tactical Air Command control in July 1976. By then the crisis in Southeast Asia had cooled down, bringing with it new constraints on the defense budget.

As early as March 1976 a private study reported that "the announced plan to break up TAC's 432nd Tactical Drone Group and put the RPVs in storage seems inevitable.

"TAC apparently looks upon the -34Ms and -34Vs as a burden, competing with the B52/Cruise Missile penetration force for the limited dollars available to them. This, despite the fact that the Loran-equipped -34Vs can lay an effective chaff corridor without requiring protective fighter cover."

(Operational doctrine then in force was to employ two piloted F-4 aircraft for each chaff mission, one to lay chaff, the other to fly cover. As a 'force multiplier,' each unmanned -34V would release two F-4 fighters for combat).

Despite presentations made to Tactical Air Command by TRA President Teck A. Wilson and program leader Starkey – and a separate suggestion that Strategic Air Command resume responsibility for RPV operations – no new funding was provided for maintaining the capabilities so well proven in the preceding 12 years.

OUR NATION'S MILITARY BUDGET is comparable to the movements of the tide – flooding when a wartime threat exists, sometimes creating tidal waves when an actual war is in progress, then ebbing when more peaceful conditions prevail. Some programs then drift out to sea on a rip tide, causing serious rethinking of budgetary priorities. Such was the situation in 1979.

TAC, for example, was forced to balance the cost of manpower, maintenance and operation of DC-130s, helicopters and RPVs themselves against the possibility of needing the capability in the immediate future in comparison with using manned aircraft in a short-term situation until the RPVs could be reactivated.

As company executives reminded the Department of Defense, the 432nd TDG was the only operational unmanned contingent available to meet any crisis. Representing a 12 year partnership between USAF and TRA, the group provided a flexible, politically acceptable system to acquire intelligence data and fly a variety of other essential missions free of risk to pilots of manned aircraft.

To no avail, members of the Armed Services Committees of the House and Senate pointed out that the BGM-34C was the "only remotely piloted vehicle program in the entire Department of Defense."

> "By this. . . decision. . .manned aircraft must now be used to fly RPV-type missions into the most hostile of enemy environments."
>
> —Sen. Strom Thurmond

TRA's appeals to those in decision-making positions within the DoD were also backed by the 432nd drone squadron's C.O., Col. James Witzel, whose entire group was eventually declared redundant. Over 60 air-launched recoverable RPVs in various configurations were then confined to the mothball fleet.

"As I left Col. Witzel's office," said TRA Vice President Starkey, "and drove out the Security Gate, I truly had an insight to General Custer's last feelings."

The full potential of Reconnaissance, Strike, ECM and Multi-Mission RPVs is yet to be realized. As Maj. Joe Tillman of the 432nd stated the case in 1978:

"WE ARE CONSTRAINED ONLY BY IMAGINATION AND MONEY."

COMPASS COPE-R

JAMES P. PORTER, U.S. Air Force

UNMANNED, UNREFUELED
RECORD ENDURANCE 28 HRS. 11 MIN. 12 SEC.

OVER THE YEARS, Ryan had developed a reputation for ahead-of-its-time aircraft design. Typical were its innovative concepts and subsequent flight demonstrations of unique high-performance manned aircraft.

These included vertical and short take-off and landing (V/STOL) research planes. Preceding the VTOLs, Ryan brought out the Navy's FR-1 Fireball and XF2R-1 Dark Shark jet-plus-propeller carrier-based fighters.

The Ryan jet VTOLs which followed were the Ryan X-13 Vertijet which could transition from a vertical take-off to normal flight, and the reverse sequence for vertical landing; and the later XV-5A/B fan-in-wing Ryan Vertifan.

Such jet VTOL research preceded by years the present operational Harrier jets of the U.S. Marine Corps, and those of the British Navy which were so effective in the Falklands War of 1982.

Similarly, Teledyne Ryan in recent years developed a number of RPV unmanned aircraft which demonstrated advanced capabilities for which no immediate mission requirement was identified by prospective military services after the advanced technology became available. This was particularly true in the case of high-altitude, long endurance designs like Compass Arrow and Compass Cope.

(TRA's **Compass Cope** vehicle is variously identified as company **Model 235**, military designation **YQM-98A** and often referred to as Cope-R.)

THE 1971 NIXON RAPPROCHEMENT with Red China had stalled additional work with Teledyne Ryan's 154 Compass Arrow vehicle, although the birds were available on call with the reconnaissance squadron at Davis-Monthan AFB in Arizona. Later, the twenty production vehicles went under the guillotine and ended up as ingots, except for their engines.

Compass Arrow's mission was high-altitude photography. Its launch aircraft had a range of 2300 miles when carrying two 154s. The 154 SPA itself could operate with a 2000-mile range, and could cruise for up to five hours, adequate for its proposed mission.

Meanwhile, a separate requirement of the military and intelligence communities was being discussed. It was for a long-endurance, high-altitude vehicle which would be self-sufficient, in that it should be **able to take off and land from a runway**, rather than being ground or air launched and parachute recovered for re-use.

The Aeronautical Systems Division (ASD) of the Air Force was interested in an R & D effort to develop a new vehicle which would serve as a 24-hour reconnaissance platform. That was a key specification requirement.

Mission capabilities would include signal intelligence gathering (SIGINT), communications relay, photo reconnaissance, battlefield surveillance, atmospheric sampling and ocean surveillance. These are the types of high-altitude 'Combat Dawn' missions flown out of Korea by TRA's 147TE and TF electronic and communications intelligence-gathering birds.

The rationale for a new vehicle type was explained by Col. Ward Hemenway, Director of the Air Force's Drone/RPV Special Programs Office (SPO) in a letter to Copley News Service:

"At its high operating altitude, this RPV would be above the jet stream, permitting it to economically loiter for extended periods of time. Its potential area of observation at high altitude would extend several hundred miles.

"Since the vehicles would be unmanned, they do not subject crews to long periods in cramped quarters; thus the manning of a unit can be more efficiently organized.

"The cost of continued operational use is expected to be quite low because the vehicle is runway-recovered after each mission. With other RPV systems, recovery has been the most critical phase. It was during this phase that most losses have occurred."

THE 'COMPASS COPE' PROJECT got its initial thrust from a proposal by Boeing Aerospace which won a sole-source contract in July 1971 to develop the system. That system, by the way, utilized the General Electric J-97 engines that were developed for Ryan's 154 Compass Arrow.

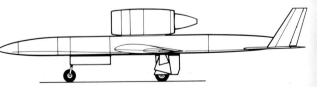

[Industry observers had speculated that perhaps the reason for the sole-source award was an Air Force effort to stimulate further competition in the RPV field in which TRA was then the traditional leader.]

A pending Air Force requirement was for a relay platform for its projected Precision Location Strike System (PLSS), a high-altitude targeting system. That was, no doubt, one of the considerations which got the Compass Cope program under way.

As an extension of its Compass Arrow experience, Teledyne Ryan proposed that it too build a 'Cope' vehicle. First studies centered on an updated 154 vehicle but that approach was soon abandoned except for some very limited applications. The tail section of the new Teledyne Ryan **Model 235**, for example, looked like that of the 154 but was larger in all dimensions.

Instead of the J-97 jet engine used in the 154, the new TRA vehicle would be powered with a Garrett ATF-3 turbo-fan engine mounted in a streamlined pod on top of the fuselage, rather than buried in the fuselage.

ARMED WITH THEIR PROPOSAL, Ryan engineers, despite a nine months late start, received a favorable response at the Aeronautical Systems Division. In June 1972, a $10 million cost-reimbursable, fixed-ceiling, no-fee contract for two test **YQM-98A** RPVs was received.

"This program," wrote Lt. Gen. James T. Stewart, ASD Commander, "like our other prototype efforts, is designed to demonstrate concept feasibility with minimum expenditure of funds before commitment is made to proceed with full system development." It was a 'fly-before-buy' contract.

"It is not expected," the General continued, "that this vehicle will satisfy the requirements of a pre-production model, [and] if the funding control cannot be accomplished, the program will automatically be terminated."

"The specification for the Cope system," explains Norman Sakamoto, veteran TRA program manager, "was equally simple. It required that we fly 24 hours; take off and land from a conventional runway with the nominal 50-foot clearances that you have in any aircraft; carry a 750-pound payload; be able to glide back in an unpowered, engine-out mode to the airfield from 100 miles away; and that we have redundancy within the systems from a reliability standpoint.

"It took us 18 months to design and produce Compass Cope-R and roll out our two orange and white birds. In itself, that was quite amazing."

THE LAST DAY OF NOVEMBER 1972, Boeing rolled out its first Compass Cope-B. Three months later TRA began construction of its two Compass Cope-R vehicles.

Eight months after roll-out, Boeing's Cope-B made its first flight at Edwards Air Force Base (EAFB). A few days later, on its second flight, the right wingtip touched the ground, the vehicle cartwheeled and was a total loss.

In January 1974, TRA rolled out its two Cope-Rs. They were built on time and within budget. Then came a financial crunch when a cut was made in the defense budget which for a time delayed both Cope-R flight testing and also affected completion of the second Boeing Cope-B entry, to replace the first which crashed.

However, Barry J. Shillito, new TRA president, remained confident that the program would get back on track. "I firmly believe this funding situation will clear up before June," he told the Aviation/Space Writers meeting in Los Angeles. Shillito also expressed concern that only 1/2 of 1% of military hardware spending was for unmanned vehicles.

He was right, at least, about Cope funding being restored.

TRAILING THE COMPETITION by many months, *Compass Cope-R (R for Ryan) was expedited through production to play catch-up.*

DRESS REHEARSAL FOR THE PRESS featured Ryan's *two birds whose super-wide wingspan was awesome.*

Bud Wolford

Robert Stewart

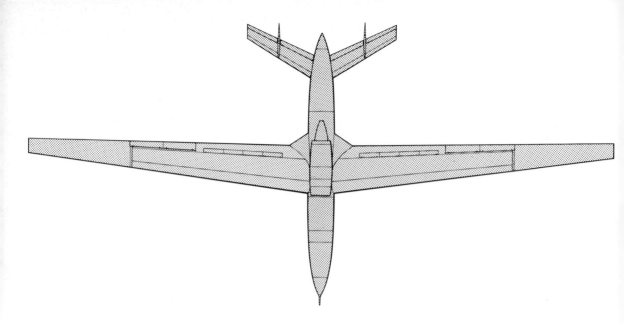

Within three months, a huge Air Force C-5A transport landed at San Diego's Lindbergh Field to pick up the two disassembled Cope-R vehicles for shipment to EAFB where they were reassembled and made ready for initial test flights.

COPE-R WAS A BIG BIRD. Wing span a super-wide 81 feet; length 37 feet and height from landing gear to engine pod, 8 feet.

Its single fuel-efficient ATF-3 turbo-fan engine produced 4050 pounds sea-level static thrust. Gross weight of the vehicle exceeded 14,000 pounds. And, Cope-R was designed to have up to 30 hours unrefueled endurance.

One design feature of the 154 which was adapted to the 235 was the slab-sided, inward-canted fuselage, an early adaptation of what later would be called stealth technology, to reduce radar reflectivity.

By August, Norm Sakamoto and his team had the two Cope birds assembled and checked out at EAFB. The No. 2 vehicle was flown off the runway August 17 for Cope-R's first flight, which lasted just 12 minutes short of two hours. Three weeks later it made a five-hour flight, followed by two more test missions in October, by which time 12-1/2 hours of total flight time was accumulated.

Meanwhile Boeing returned to EAFB with its second YQM-94 Cope-B which made its first flight November second. Earlier, the Air Force had provided some funds for building Boeing's second vehicle. Col. Hemenway had pointed out that the Compass Cope program was to be a technology demonstration and not a 'fly-off'. After demonstrations, an orderly engineering development program would have to be initiated.

To make life more interesting in the desert test area, the Northrop B-1 supersonic bomber was due for its roll-out at nearby Palmdale, and fighter and attack plane competitions were ongoing at Edwards.

WITH JAWS LIKE A HUGE SHARK, an Air Force C-5A *ingests the dismantled 'Copes' for transport to Edwards Air Force Base.*

Robert Stewart

AIRBORNE FOR THE FIRST TIME, the Model 235 *responds to the experienced touch of ground controllers, recording nearly two hours of initial flight test time.*

Bud Wolford

WITH FOUR GOOD COPE-R flights chalked up, TRA was ready to go for a long-endurance test. The crew got the bird off the runway early Sunday morning November third.

Sakamoto picks up the story: "We rotated our technical people four hours on, four hours off. We had two MCGS (microwave command guidance system) vans plus a trailer out on the lake bed next to the command station. One van was for back-up in case the other went down. We didn't have any of our controllers on the runway as they would be needed there only when we were ready to come in and land.

"We flew right on through Sunday night and into Monday, the day the B-1 was being

EDWARDS' DESERT MOONSCAPE with parched terrain *was the locale for the white-knuckle check-out before the first flight.*

Ed Precourt

rolled out. By then an Air Force F-15 wanted to fly a mission at Edwards so they called the Director of Operations, Col. Jim Woods, who was then at the B-1 roll-out. They asked that the Cope-R come in and land so they could get on with their mission.

"Woods' reply was, 'If those guys at Teledyne have the guts to stay up all night and they are still flying, you and your F-15 just went to the bottom of the priority list. They can stay up there as long as they want to.' So, we continued flying until about noon.

"Our controllers — Buck Weaver, Ed Sly, Art Rutherford and Chuck McGean — were doing a superb job. Buck, who had experience as a take-off and landing controller of Navy manned aircraft converted to drones; and veteran Ed Sly, were our lead controllers. For the overnight flight we had trained a few other people in case we had an unusual problem, which fortunately we didn't.

"By the time the flight was over, I'd been up 45 hours so needed some shut-eye while things were quiet in the middle of the night. I decided to stretch out on six chairs in the telemetry area at the command station. I had just started to doze off when I heard 'engine run-down' at 1:00 a.m.

"I dashed into the command room and lo and behold, the engine rpm gauge had dropped down to nearly zero. I got on the horn and called Ed Sly. 'What do you show?' Ed's gauge didn't show any rpm decrease but the Garrett engine rep requested that we increase our rpm slightly. We did and all the bird did was climb a little higher. We finally determined that one of the sensors had malfunctioned. After that, no one could go back to sleep. At Air Force request we beeped the rpm back a bit and the bird descended to its earlier cruising altitude."

113

FOUR HOURS ON, FOUR HOURS OFF, Norm Sakamoto *(second from left) and the controller crew maintain the vigil for the record-breaking unmanned flight.*

"**A**FTER ABOUT 26 HOURS FLIGHT, my boss, Bob Reichardt, came into the control room and whispered in my ear. 'The world's unrefueled record for unmanned vehicles is 27.9 hours [flown by a turbo-prop-powered converted sailplane]. Is there anything we can do to break that record?' I said, 'Sure. Just fly.'

"Quietly, Reichardt is telling me this with an Air Force officer sitting next to me on the other side. I doubt he was interested in our breaking any world's records, but we felt it would certainly be a worthwhile achievement. I got on the phone and called Buck Weaver out on the lake bed.

"Buck agreed we could break the record 'if we have enough fuel.' I told him my reading of the fuel gauge was that we had enough for 30 hours when we took off, and at 26 hours we could still stay up perhaps another four hours.

"At 27-1/2 hours, we were making our final descent over the Tehachapi-Barstow area towards the south Rogers Lake bed at Edwards. Buck recognized about where we were relative to the record, so instead of landing, he does a fly-by at about 1000 feet over the runway; goes down to the end, makes a left turn, goes back out to the range and comes back in to land.

"Meantime, the Air Force calls Buck on the horn. 'What was the reason for your go-around?' Buck's response, 'I saw some data on my control panel I wasn't happy with, so I wanted to check it out before we attempted a landing.'

"The landing was okay. Flight time: 28 hours, 11 minutes, 12 seconds. A new world's record for unmanned, unrefueled flight!"

Three weeks later, Boeing's No. 2 bird racked up a 17 hour 24 minute flight, culminating in the first RPV night runway landing.

As to the Cope-R record, an Air Force spokesman revealed that when the YQM-98A took off, maximum power was used. The Gar-rett turbo-fan was then "throttled back to intermediate and a maximum continuous cruise/climb was initiated, resulting in gradual altitude gain [possibly to around 70,000 feet] as fuel was burned off."

FACED WITH POSSIBLE INCLEMENT winter weather in the California desert, and to make room at Edwards for higher priority flight test operations, the decision was made to move Cope-R operations to the Eastern Test Range, based at Patrick AFB in Florida. At Edwards, Cope test flights had been restricted to weekends.

A second advantage would be gained: how well could flights of a runway-based unmanned vehicle be integrated with FAA air traffic control of commercial airline flights?

There was also speculation in the trade press that the Cope-R vehicles might be "prepared for reconnaissance of Cuba and surrounding areas. These reports, however, could not be confirmed."

COMPASS COPE-R's LANDING was uneventful *and picture-perfect after setting a new record of 28 hours, 11 minutes for unrefueled drone flight. Proud crew gets the credit — and a picture to prove it.*

114

ESCAPING ANTICIPATED BAD WEATHER and out-ranked *by priority manned flights, the two Cope birds were airlifted to Cape Canaveral, Florida.*

There was another side of the move-to-Florida story as related by Norm Sakamoto:

"Lt. Col. Richard Wright of the Systems Command at Andrews AFB, was visiting our Cope-R operations at Edwards. Earlier he had been assigned at Ryan, San Diego, during our Vietnam spy-plane days. While at EAFB he said to me, 'You know, we ought to put a payload in your Cope-R and go fly somewhere.'

"My reply was that we certainly could, and suggested we take the Combat Dawn electronic intelligence payload we used in the 147TE flights from South Korea and install it because the Cope payload compartment was designed that way. We incorporated the idea in a formal proposal which the Air Force accepted, and we did the preliminary work while finishing up the Edwards activity."

In mid-February 1975, the two Cope-R birds were airlifted by C-5A Galaxy drooped-winged transport to Florida for operations at Cape Canaveral.

Further flight testing there would provide additional evaluation of the Tactical Landing Approach Radar (TALAR IV) system and expand the operating envelope of the Garrett engine at altitudes over 50,000 feet.

Meantime, upon completion of the TRA and Boeing demonstration flights at EAFB, the Boeing YQM-94 vehicle had been returned to Seattle.

FLIGHT TESTING RESUMED at Cape Canaveral May 8, 1975 and continued through September during which time 12 more missions were flown. RPV No. 2 made its sixth and seventh flights in Florida, racking up a cumulative total of 48 hours. Then No. 1 bird flew the remaining 10 flights, bringing total Cope-R flight time up to 90 hours.

Tactical Air Command, which the next summer assumed responsibility for all operational RPVs, visited Patrick AFB during the test program and showed some interest in acquiring several vehicles to determine whether or not TAC could generate a requirement. Nothing solid, however, came of the inquiry.

"During our 12 flights in Florida," Sakamoto explained, "we — that is, the Air Force and TRA — acquired a lot of data including some from that little island nation south of Florida. That type of intelligence data was said to have been quite interesting, but we didn't get involved in its interpretation which was for the purposes of government agencies, not ours.

"We were a very proud, professional group. When we had VIPs visiting, our people were all dressed in white jumper-type outfits. Bill Evans, a General in Systems Command, commented that 'I have never seen a more professional looking group of technicians, nor an organization more smoothly operated, even in the military.' It was a generous comment and a real morale booster."

THE COMPASS COPE TESTS involved a whole new area of operation for the contractor. The only previous experience with runway takeoffs and landings of unmanned vehicles had been racked up by the military services.

TRA's Ed Sly, responsible for control of Cope-R after it was airborne and clear of the ground environment, had more hours of remote control than any other technician in the business.

However, finding the right man to handle guiding the broad-winged (81 feet) RPV down

AT EVERY STEP IN THE COPE PROGRAM, *the best technicians available groomed the vehicles for the rigorous test flights.*

DAYTIME LANDING AT CAPE CANAVERAL. At night, in a
terrible rain and thunderstorm, it was a different story.

the runway on takeoff and return safely to the
landing strip was solved by selecting one of
TRA's engineers already on the payroll.

Keith B. (Buck) Weaver had joined the company in 1964, following retirement as a Navy
Commander and jet-qualified fighter pilot.
While at the Atlantic Fleet Missile Range at
Roosevelt Roads, before the days of supersonic
Firebee target drones, Buck saw the phaseout of surplus F9F fighters which were then
droned for air-to-air weapons exercises.

Buck became one of the best controllers,
guiding the unmanned fighters down the runway on takeoff, then returning them to the
landing strip (if the F9F as a target was still
flyable) where, as ground controller, he could
visually guide the jet from a 'Fox Cart' to a
normal landing. With hundreds of missions
under his belt, Buck was well-qualified for
similar assignment on Cope-R.

"On the September 29-30 nighttime mission," recalls Sakamoto, "I was on the phone
tracking the flight from my mother's home in
San Jose, California. Ed Sly was at the control station, keeping me informed about the
weather.

"It had rained four inches in one hour while
we were in the air during which we lost a
generator and a back-up electrical system.
Although our systems were redundant and
therefore we didn't have a crisis, we decided
to play it safe and come home. At one a.m. we
were flying through a terrible rain and thunder storm coming in to land.

"The runway control system was not designed for that kind of storm conditions. Buck
Weaver, of course, was at the controls for the
landing sequence. Unfortunately, the trim

bias had not been changed under those stressful conditions to a zero indication.

"Buck had about one degree in there and
that one degree was sufficient to cause the
bird to land with one of its main wheels off the
200-foot-wide runway where the soil was very,
very soft. The Cope-R cartwheeled, stripping
the other landing gears off. The resulting fire
was put out by the base fire department. The
plane was not re-built, since the test program
was essentially completed.

"TRA cooperated fully with the Air Force in
the subsequent investigation. It was very
open, very honest, well-documented and
closed in short order, to the customer's complete satisfaction. The conclusion was that
there were no problems with the basic system."

As GENERAL STEWART of the Aeronautical Systems Division had pointed out when
awarding the contract for the YQM-98A, there
was no expectation that the vehicle would be
other than a demonstration model.

It was implied that if successful, a further
study of High-Altitude, Long-Endurance
(HALE) RPVs would follow. With that in
mind, the Air Force, in December 1974, contracted with both Teledyne Ryan and Boeing
for system engineering studies to aid the military in deciding whether or not to proceed
with full-scale development of an advanced
HALE vehicle.

"Both companies," Sakamoto explained,
"made investigative studies which included
periodic oral presentations to ASD. The way
it was structured, we sat in on Boeing's briefings and they on ours. The Air Force wanted
to get the best, most experienced engineers
working together, sharing potential solutions
so the ultimate requirement, the request for
proposal for a HALE vehicle, would take advantage of this combined intelligence."

In recognition of "the magnitude of the
Cope program," Barry J. Shillito, TRA President, named Vice President Erich Oemcke,
"the single most qualified person," to assume
the responsibility of Cope program director.

Meanwhile, TRA was flying its final tests of
Cope-R at Cape Canaveral. These concluded
the end of September 1975. In April, 1976,
ASD issued to TRA, Boeing and ten other
companies its Requests for Proposal (RFPs)
"to select a single contractor for pre-production prototype development of the Compass
Cope HALE system." The contract would be
for three vehicles plus support equipment and
facilities.

Proposals were due in June with award to
a single contractor, according to an ASD press
release, expected in the fall.

TRA'S PROPOSAL was for an advanced version of its Cope-R Model 235, having company identification **Model 275**.

"We changed the configuration," Sakamoto recalls, "to be in line with the RFP requirements for a modular concept for maintenance and serviceability; and the power plant was switched to the General Electric TF-34 engine. This engine was also used in the A-10 attack aircraft and would provide the low specific fuel consumption needed for long endurance flight. It gave us 7500 pounds of thrust compared to the 4050 pounds of the 235 Cope-R vehicle.

"We also proposed a streamlined, faired-in, submerged engine installation, with better air flow than previously provided with a top-fuselage-mounted engine and pod.

PROPOSED MODEL 275 ADVANCED VERSION in artist's *concept, below, shows changes in configuration from Model 235 pictured above in flight.*

"Special emphasis was given to the avionics/control systems because redundancy there is the best assurance of achieving reliable long-endurance flights. Designed-in, back-up reliabilities comparable to those for manned aircraft were a major consideration in the 275 proposal.

"Easy, eye-level access for maintenance of all systems was provided, thus eliminating work stands and overhead hoists."

With the new requirements specified by the RFP and contractor design updating, the new Compass Cope pre-production vehicles would be capable of carrying a variety of larger and/or heavier payloads and would have more advanced avionics and landing systems. They would perform many different missions with a minimum of configuration changes, and would operate at altitudes above 55,000 feet for up to 24 hours, as YQM-98A had already done.

THE CONTRACT AWARD was not long in coming.

On August 27, 1976, a Friday, the Air Force announced that Boeing Aerospace had been selected as the single contractor for pre-production prototype development of the new HALE vehicle. Of an estimated $77.2 million award, $2.75 million was to be made available immediately toward building three pre-production units.

Teledyne Ryan Aeronautical cried 'foul,' and immediately protested the sudden award to the General Accounting Office (GAO). The sudden award aspect was due in part to the fact that General Stewart was to retire Wednesday, September 1st. Also, TRA's 'best and final' offer, already prepared, was due September 3rd.

Although the 'competition' was not a 'fly-off,' it appeared that sufficient consideration may not have been given to the comparative flight tests of the Cope-R and Cope-B vehicles.

Ryan had flown 90 test flight hours in 17 missions at two locations. Boeing had only three flights at Edwards AFB. Both contractors had lost one of their two birds, Boeing on their first flight, Ryan on the last of their 17 flights. Ryan's longest flight endurance was 28 hours; Boeing's 17.

TRA's protest made specific mention of an appearance that technical preference may have been given for features not contained in the RFP, and that Boeing's cost estimates were based on 'probable' costs. Many felt that the award should have been judged on parallel submissions to a precise RFP without variations.

The rationale for moving the award date up appeared to have been that General Stewart was on the source selection board and that funds for the first increment of $2.75 million were then available.

Press reports six months after the contract award indicated that GAO reported its investigation had been completed February 1st and that the file was being studied. GAO was also reported to have asked the Defense Department to review its requirement for the new vehicle.

As they say in the vernacular, "Compass Cope appeared to have died on the vine." However, the technology gained in high-altitude long-endurance (HALE) flight is still being applied.

DAWN AT EDWARDS AFB *sharply outlines the sleek profile of Compass Cope-R. It also forecast the later sunset of this record-setting high altitude, long endurance UAV.*

FIREBRAND

Composite Photo

Anti-Ship Missile Target

*AMONG THE RAPIDLY GROWING
threats to the security of U.S. Navy ships in the
mid-1970s were Russian cruise missiles.*

*The concern of Navy planners dated back to
the Six-Day War of June 1967 between Israel
and Egypt. The United Nations-arranged
cease fire had been broken that October when
Egyptian torpedo boats sunk the Israeli destroyer 'Elath' with Russian-made 'Styx' missiles coming in at low altitude off Port Said.*

*Since that encounter deadly new sea-skim-
ming cruise weapons such as the SS-N-9 Siren
Mach 1.4 surface-to-surface missile and the
AS-3 Kangaroo Mach 2 air-to-air projectile
became formidable opponents challenging
U.S. defense systems.*

FIREBRAND

NEW EFFORTS had to be made by the
Navy to procure an Anti-Ship Missile Target
(ASMT) capable of being threat-replicative
— a system that could fly at supersonic
speeds and at both high and low altitudes.
The turbine-powered ZBQ-90A had been cancelled in 1973. Although capable of meeting
most Navy requirements, its development
costs exceeded the budget available.

In 1975 the decision was made to develop
two target systems to simulate the anti-ship
missile threat. Bendix was contracted to
build the MQM-8G Vandal, a converted RIM-
8G Talos missile, to duplicate the terminal
phase (final 24 nautical miles) of the enemy
missile approach.

Teledyne Ryan was chosen to head the team
which would design and develop a new vehicle
to simulate the mid- and low-altitude threats
throughout the total mission profile of an op-
posing cruise missile.

The new design was to use currently available technology in airframe, propulsion and
avionics. Unlike the Talos missile target, the
Ryan target would be recoverable. It would
also have to be usable against any of the shipboard defense systems to be available by the
mid to late 1980s. Mentioned were such defense weapons as Sea Sparrow, Aegis and the
Close-In Weapons System (CIWS), as well as
5- and 8-inch guided projectiles and high energy laser (HEL) weapons. The new target
would be used only in weapons development,
not training.

TRA had already developed the Supersonic
Firebee II BQM-34E/F and had a radar altimeter low-altitude control system (RALACS)
with which Firebee targets could be flown as
low as 50 feet above the water at speeds of 500
miles an hour. The Navy had already been
using the BQM-34E (Mach 1.1 @ S.L.) as its
supersonic target at Pt. Mugu.

A FORMAL CONTRACT for the **ZBQM-
111A 'Firebrand'** was awarded in May 1977
by the Naval Air System Command. Test of
the target system was expected to lead to an
initial production contract for 50 systems.
Initially, however, nine of the targets would
be built — six for R&D and three as special
test vehicles.

There had been some conflicting opinions
within the Navy regarding fiscal aspects of
the program. It was known that there were
not sufficient funds available in the budget to
take the program through to completion.

However, Naval Operations said they understood, but to proceed anyway. And, the
target acquisition office at the Assistant Secretary level still felt the Navy could get by
with the Talos missile conversion.

In any case, it was believed the Firebrand
could be operational in 1983 at Pt. Mugu and
perhaps at the Navy Atlantic Fleet Weapons
Test Facility at Roosevelt Roads, Puerto Rico.
Program manager for Firebrand was Sam
Sevelson who had performed the same role for
Ryan on development of Firebee II.

The basic Firebrand design called for twin
ramjet engines for sustained flight and a
booster-engine assist during the launch phase
of operation for either air or ground launch.
Wing span would be a scant nine feet and
length 34 feet.

Ramjet engines were selected as the most
economical propulsion units to meet the speed
and range requirements at very low altitude
operations. Thiokol would provide modified
Patriot surface-to-air missile rocket motors as
booster units.

Cdr. Eugene E. Auerbach, project manager
for the Navy, was an enthusiastic supporter of
the Firebrand program. "This is kind of a new
concept," he stated. "Instead of designing it
to meet a particular weapon system's need,
we're designing this vehicle to be threat repli-
cative and fly against a variety of weapon
systems." The prime objective of the bird was

WIND TUNNEL TESTING determined performance of Firebrand design with 'Patriot' rocket booster to assist either air or
ground launch.

to evaluate the acquisition, tracking and intercept capabilities of the Navy's defensive weapon systems.

The ZBQM-111A (TRA **model 258**) was one of the sexiest, meanest and deadliest looking targets ever designed by the company. With a body diameter of only 28 inches, with stubby aft swept delta wings and dual ramjets snuggled below the elevons, it even looked supersonic while sitting on ground handling equipment.

NAVY PROCUREMENT POLICY for the contract, as described by Cdr. Auerbach, stated, "We've basically gone to pulling off-the-shelf systems and integrating them in order to gain the reliability that's been developed in other weapon systems, and gain a price break by joining with those other projects that are further down the learning curve."

Marquardt was selected to develop performance characteristics for the ramjet, and was contracted to build ten shipsets. Each engine would develop about 5,000 pounds of thrust with normal shock inlets and be designed for a three mission life cycle. The decision to use two externally mounted ramjets instead of one large engine was based on cost and the

technical problems associated with off-center thrust of a single engine large enough to meet the power requirements of the Firebrand.

The Thiokol boosters were to serve two purposes, depending upon mission profile selected. Normally used for ground launch, the booster, attached to the bird, could be carried aloft for air launch, supplementing the ramjet engines to accelerate the Firebrand to Mach 1.9 following drop from the mother ship. During ground launch the booster would accelerate the target to Mach 1.3. The ramjets were designed to take over propulsion at Mach 1.2.

Among other off-the-shelf items considered was a control system module of the Navy's Control Data AN/AYK-14 flight computer used in the F/A-18 fighter. The Honeywell 7193 radar altimeter used in the Tomahawk cruise missile and in the Harpoon missile was selected for integration into the new target. As prime contractor and integrator of the support systems, TRA's staff of engineers, liaison personnel and Washington contacts were taxed to capacity.

Wind tunnel tests of a one-seventh scale model at NASA's Ames Research Center received nationwide coverage in the leading aviation publications. The Firebrand seemed well on its way to becoming one of the Navy's outstanding missile threat simulators.

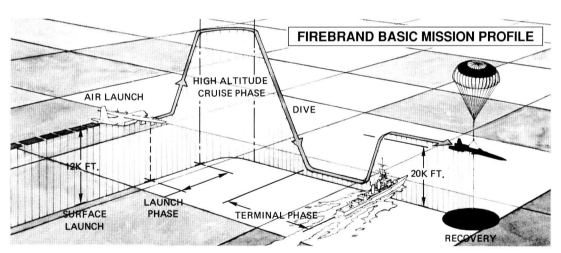

FIREBRAND BASIC MISSION PROFILE

AIR LAUNCH

HIGH ALTITUDE CRUISE PHASE

DIVE

12K FT.

20K FT.

SURFACE LAUNCH

LAUNCH PHASE

TERMINAL PHASE

RECOVERY

PROFILE FOR A TYPICAL Firebrand mission would begin with a booster assisted air launch from a DC-130 to accelerate the ZBQM-111A to supersonic speeds before the ramjets take over propulsion. This would take the target to 40,000 feet for the Mach 2 cruise phase followed by a dive to 300 feet for the final run toward the defending ship.

Firebrand needed to simulate both air-to-surface and surface-to-surface missile attacks such as being launched from an enemy air-

craft at high altitude or launched from a submarine. This demanded that a very wide operational envelope be met.

In all cases, the terminal portion of the mission required the target fly in the general direction of a defending surface ship.

"Now," explained Norman Sakamoto, "no skipper in his right mind would want a larger target, even like the Firebrand, to be coming direct at him. Just the energy of the mass would be damaging, much less the explosion

FROM MACH 2 AT 40,000 FEET to deadly attack at 300 feet, *the ZBQM-111A was designed to challenge the Navy's best defense systems.*

that would occur because of the fuel-air mixture in the fuel tank.

"So we used a TACAN terminal guidance system that would cause Firebrand to fly an offset of a thousand yards or more. After the simulated attack Firebrand would pop up before it reached the ship, slow to subsonic speed and deploy its drag parachute and main chute system for a water recovery.

During the terminal phase of the mission, the tactical air navigation unit would lock onto a homing beacon on a platform being towed by a ship.

"Theoretically," said Cdr. Auerbach, "sophistication of the [on-board electronic equipment] should be good enough to keep the target right on the offset course. However, we want to make sure because we're going to fly it at a manned ship."

Because of the fast closure rate toward the ship a fail-safe system to exit the Firebrand was an essential consideration. Recovery would be automatically initiated should Firebrand deviate from the pre-programmed flight envelope.

Since the whole project was a new attempt to realistically simulate an enemy missile, the design tasks were complicated by customer-dictated range safety measures which eventually resulted in a redundancy of gadgetry.

IT WASN'T UNTIL JANUARY of '79, following two years of design, redesign, full-scale mockup, component integration and wind tunnel tests that the first 'flight' was made in the TRA flight simulation laboratory. In the course of development, engineers performed more than 1,000 hours of

TWO YEARS OF RESEARCH preceded any potential flight test *with design, full-scale mockup, component integration and wind tunnel testing.*

real-time Mach 2.1 fifty-foot altitude flight simulation on the program. By then the factory was beginning to fabricate parts, and the first actual flight was a year away.

It was at this point that the Navy called for time out to take a closer look at monies already spent, budget status and cost to complete fifty operable units. At the urging of Richard Rumpf, Navy Director for Air & ASW Programs, the Vice Chief of Naval Operations agreed to convene a Defense Resources Board meeting over the Christmas holidays at which the program was put on hold.

Despite the fact that the Firebrand design, engineering concept and wind tunnel tests met or exceeded the Navy requirements, the decision was made to halt further development and future flight testing.

In addition to basic funding problems, several other factors were responsible for 'noncontinuance' of the program. For one thing, Firebrand, with its dual ramjets and booster, was a heavy bird, perhaps too large a load for the DC-130 or for ground launch. Possibly B-52s might be required as launch aircraft. That raised the question of additional funding and launch aircraft availability.

THE ENTIRE PROGRAM provided a useful case of history lessons to be learned on developmental projects requiring extensive precontract study and assured funding to completion. The contractor, subcontractors and supplier all had strong reasons to stress the capabilities of their own products and customer-dictated requirements had to be given priority. Some proved to be impractical.

Nonetheless, the design and development exercise provided the Navy and TRA with invaluable experience for future requirements.

It created a rich background in new target technology, including the use of sophisticated digital avionics control systems.

Aware that there were still 500 surplus Talos missiles available, the decision was to continue use of MQM-8G Vandal targets for the terminal phase of missile simulation for evaluation of new weapons.

The stage was now set for further development in the cruise missile target field, and eventual participation in the Air Force's High Altitude High Speed Target (HAHST) program. Almost simultaneously an additional requirement for a supersonic low-altitude target (SLAT) was being considered. Experience gained with Firebrand gave TRA confidence in submitting their proposal for the new sea-skimmer system.

GROUND LAUNCH CAPABILITY with rocket booster *attached was checked for size and fit at Navy's Pt. Mugu facility.*

U.S. Navy

Dave Gossett

GROUNDED AGAINST SAN DIEGO skyline, *full scale Firebrand program was held back by budget constraints.*

TRA TARGET/SPEED ALTITUDE ENVELOPES

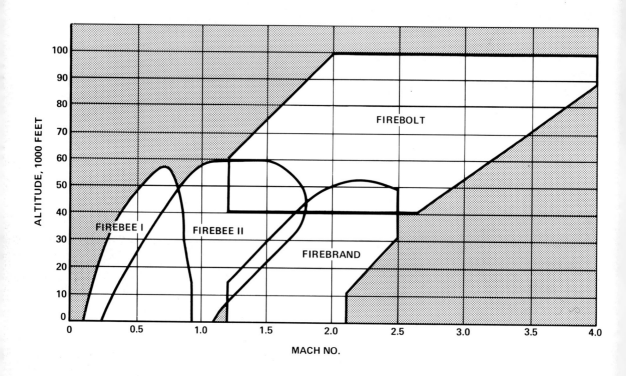

FIREBOLT

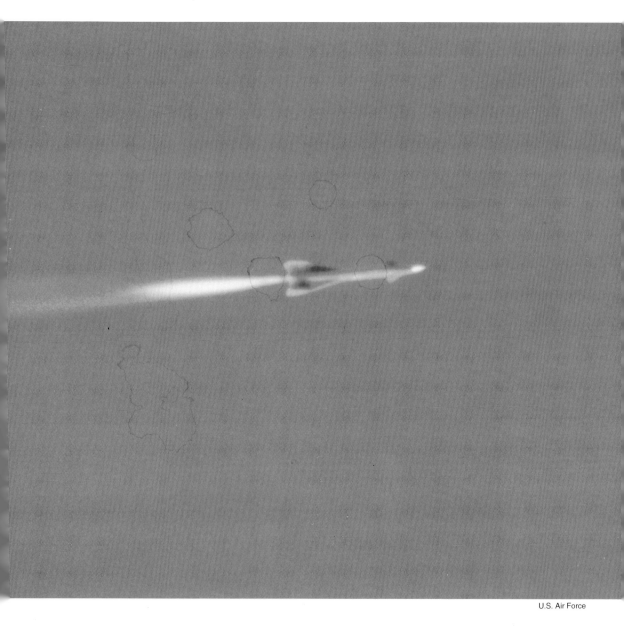

U.S. Air Force

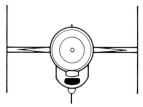 MACH 4.3/103,000 Feet

125

To MAINTAIN OPERATIONAL readiness of its fleet of aircraft carriers, the Navy conducts 'underway fleet training.' To evaluate effectiveness of its capability to defend carriers against attack, the Navy sent out a request to industry for an expendable, 'throw-away' aerial target which could be launched from a carrier-based plane.

The advantage of the expendable target is that the carrier, during training exercises, need not stop to pick up a 'recoverable, parachute-for-reuse target' from the open ocean; then decontaminate the target and get it back into operational condition.

THE GROWTH OF PROPULSION systems for unmanned aircraft reflects the tremendous strides which have been accomplished in recent decades. From tiny gasoline powered model airplane engines to sophisticated devices producing thousands of pounds of thrust, the history of aircraft advancement is clearly demonstrated.

In 1903 the Wright brothers designed and hand-carved their first propellers. Gasoline engines propelled the aircraft for several decades. The advent of the jet engine, developed during World War II, revolutionized the industry, and turbo-props supplemented jets in many designs. Rockets were developed to aid in providing additional boost in numerous applications.

Ryan pioneered the jet-powered target field as early as 1951, and the thousands of RPVs produced by them followed the continued improvement in jet engines, starting with the subsonic Flader XJ-55 jet planned for the XQ-2 Firebee, up through the advanced supersonic Teledyne CAE engines.

Departure from the jet concept to the propulsion system of the **Firebolt AQM-81A/N** was a major step toward attaining target altitudes exceeding 100,000 feet and speeds of over Mach 4.

Earlier the U.S. Navy had issued an RFP to industry for an expendable throw-away target to be used for underway fleet training. Launched from carrier-based aircraft, the target, equipped with a rocket engine, was to be used for fighter-intercept flights in open sea ranges.

The company which won that competition is still in production of its AQM-37 expendable target. It is powered by a liquid rocket engine using nitric acid as an oxidizer and hydrazine as the fuel. When the two liquids come in contact they start burning without benefit of an ignitor.

The engine is technically known as a hypergolic rocket system. Needless to say, the AQM-37 propulsion system is treated with much respect.

THE AIR FORCE later expressed interest in a similar but parachute-recoverable target having greater payload, higher altitude capability, a safer propulsion system; and a higher speed of Mach 4.

Teaming up with the builder of the AQM-37 and the Chemical Systems Division of United Technology, the Air Force contracted for a hybrid propulsion system capable of meeting the new requirements. 'Proof of concept' was demonstrated through a limited flight test program.

Larry Emison, later program manager on the Firebolt program, described TRA's entry into the Air Force competition:

"Our competitor almost had the contract sewed up tight and submitted its proposal prior to a call for competitive bids by the Air Force in early 1979.

"When the original manufacturer's costs appeared to be excessive, the Air Force issued a competitive RFP to industry. Frank Oldfield, TRA Director of Engineering, persuaded then President of TRA, Teck Wilson, to go all out on the program."

In December, USAF's Armament Development Center at Eglin AFB, Florida, announced award of a contract to TRA for full scale engineering development of the High Altitude High Speed Target (HAHST). (The earlier designation for the Mach 4 target was HAST — High Altitude Supersonic Target).

TRA's in-house designation for its proposed system was Firebolt; after contract award the military nomenclature became AQM-81A for the Air Force, and AQM-81N for the Navy.

LETHAL APPEARANCE of Firebolt *is evident even as it rests on its transportation dolly.*

A TECHNICAL DESCRIPTION of the new hybrid rocket engine reads: "The controlled thrust assembly contains a solid fuel grain compound of butyl rubber and Plexiglass. This fuel is totally inert and requires very precise combination of vaporized oxidizer gases and ignitor to start combustion.

The liquid oxidizer, inhibited red fuming

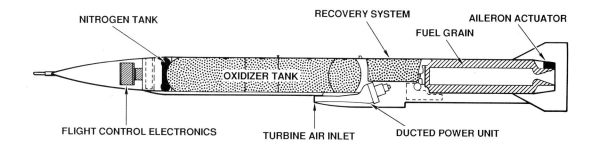

NITROGEN TANK · RECOVERY SYSTEM · AILERON ACTUATOR · FUEL GRAIN · OXIDIZER TANK · FLIGHT CONTROL ELECTRONICS · TURBINE AIR INLET · DUCTED POWER UNIT

DETAILS OF PROPULSION SYSTEM are revealed *in cut-away drawing which also shows other systems.*

nitric acid, is pressurized by gaseous nitrogen. This pressure transfers oxidizer from the storage tank to a ram air-driven high pressure pump. The oxidizer, under high pressure, is delivered to the oxidizer manifold via the throttle, where it is injected into the thrust chamber."

Emison continued: "The biggest advantage of the new engine over the earlier liquid rocket engine was the ability to change the thrust for varying altitude and speed requirements. Understanding the principle of its operation was relatively simple, but far beyond the comprehension of a shade-tree mechanic!

"Package this mouthful of razzle-dazzle into a tube 120 inches long and about a foot in diameter and you have 1,200 pounds of thrust capable of pushing Firebolt through the sky at unbelievable speeds and altitudes."

Even while resting on its transportation dolly, the Firebolt had a lethal appearance. With its diminutive canards well forward on the fuselage, with streamlined vertical fins enclosing the ailerons, it looked more like a formidable cruise missile, rather than a target aircraft. Given an explosive warhead, Firebolt had the possibility of performing a dual role as target or as a missile.

Fully controllable throughout its performance envelope by either pre-programmed automatic profile control or command by a remote pilot, it could make speed, altitude and heading changes throughout its flight performance envelope.

A Mach 3.5 mission flown at 90,000 feet could achieve cruise ranges is excess of 90 nautical miles, providing ample opportunity for multiple missile attacks by several shooters. Had it been designated as a cruise missile, its survivability against any anti-missile weapon would have been very good.

FIREBOLT IS IDEALLY SUITED for open ocean and forward deployment area operations since it is launched from tactical aircraft and carries its own scoring capability. At the conclusion of flight, the target

and its payload is recovered by a two-stage parachute system. This assures reuse following either mid-air retrieval or surface recovery after letdown.

In the highly competitive RPV business, the eventual receipt of a contract is due, in large part, to the skill and dedication of the men and women who prepare the proposal. In a hectic 90-day period, a team, under Dave Campbell, working around the clock, seven days a week, submitted their proposal to Eglin Air Force Base in mid-1979. After a more thorough study of the requirements, a resubmittal was sent to Eglin, and in December, TRA received a Christmas go-ahead.

July 6, 1981, was a decisive date if the program was to continue. The Critical Design Review board of the Air Force spent a week at the plant and at the conclusion of the review Col. Fred Carnes, leader of the team remarked, "This meeting was the most successful design review in which I've ever participated." By September nine targets were being produced for the flight test program to begin at Eglin in early 1983.

Logistics for the move from San Diego to Eglin was a simple matter of crating the targets in regular shipping containers and sending them by common carrier. Check-out of the birds on a relatively uncomplicated console was accomplished, as was check-out of the launch equipment in the Air Force A4-D fighter aircraft.

Instead of the sophisticated console required for air launch in the DC-130s used for larger birds, the fighter jock in the A4-D had only two buttons to push to activate and launch the target.

THE FIRST SUCCESSFUL flight test of the **Model 305** Firebolt target was made on June 13, 1983 at the Eglin, Florida gulf test range. After launch from the A4-D at 50,000 feet and Mach 1.5, the bird executed several high-G maneuvers at Mach 2, ten thousand feet higher than the launch plane. The second mission on July 20 was a repeat perfor-

EAGER TO FLY, the Model 305 made its debut *on schedule, with a successful first flight June 1983.*

mance. Unlike the expendable AQM-37, both targets were helicopter recovered for re-use.

One of the least publicized aspects of the development program was the award to the Eglin test group in January 1984, from the National Aeronautics Association, of two plaques commemorating the Firebolt's world record for unmanned sustained level flight, both for speed and altitude.

Flying at 103,000 feet and Mach 4.3, the stagnation temperature blackened the stainless steel leading edges of the airframe. However, Air Force studies proved that ablative materials, such as used on the space shuttle, were not required.

By June of 1984 the Development, Test and Evaluation flights were completed. (16 missions were flown and only one loss occurred and that because of a planned direct missile hit.)

The next phase following the development flights was the IOT&E phase (Initial Operational Test & Evaluation). Under the leadership of veteran Jack Young, assisted by Ed Sly and other seasoned TRA technicians, the Eglin operation was completed in the fall of 1984.

Shortly after receiving the Air Force contract, TRA was the recipient of an additional order from the Navy for 12 birds. Several flights were flown at Pt. Mugu in July, 1984. Two of the six targets were expended during this period. On one flight it was reported that the pilot carrying the target to altitude had a choice of two buttons to push, and inadvertently punched 'launch' rather than 'activate'. A pre-flight checkout on another bird resulted in a launch error, and the AQM-81N established a world's record high dive into the Pacific.

FOR THE PAST SEVERAL DECADES, the procurement problem for all U.S. military services who have requirements for pilotless target aircraft has been one of expendable vs. recoverable PTAs.

Many factors are involved in the procure-

ment decision. The simplest question posed is "How much is the unit cost?"

Taking a hypothetical, uncomplicated case, assume that the 'throw-away' drone sells for $20,000 each, and the recoverable drone sells for $100,000. The throw-away target makes one flight against any shooter — no recovery costs with parachutes, helicopters or replacement parts are involved. Ten missions using an expendable at $20,000 per flight amounts to $200,000. A recoverable target selling for $100,000 and capable of sophisticated control and scoring systems, and the ability to survive ten flights, gives the user a much more economical bird, even considering the turnaround costs.

How many flights can the recoverable target make against the weapons fired? Flights per target vary considerably depending upon the training or combat environment to which they are exposed. In Vietnam, some recoverable RPVs flew as many as 68 missions in a hostile environment, and others were shot down on the first mission.

The average for Firebee targets flown for the Navy firings at Roosevelt Roads against missile ships and fighter planes was more than 12 flights per bird. The same average was experienced earlier with Army firings at White Sands and McGregor ranges.

Granddaddy of all Firebee targets with the most flights was 'Old Red', a BQM-34A used by the Air Force at Tyndall AFB, Florida. It held a record of 82 flights spanning five years and three months of service.

Next, is the capability of the target to simulate the offensive weapon against which air or ground fire is directed. A simple comparison could be using clay pigeons for target range firing vs. shooting at live doves or ducks. Preprogrammed shooting in a controlled, predictable environment differs greatly from firing at a live bird in the open.

To a major extent the similarity between a clay pigeon and a live bird accounts for the difference in cost for types of aerial targets. To more closely simulate the threat, expensive equipment such as IR/visual/radar augmentation, scoring systems, high 'G' maneuvering equipment, ground or air control station and launching costs, either ground or air, must be taken into consideration.

THE PRIME OBJECTIVE of using unmanned targets is to train the shooters to execute their missions in the most realistic environment obtainable. At first glance, to acquire that realistic simulation is costly, if comparing expendable against reusable targets.

The Defense Department spends millions of dollars annually in the research, test and development of unmanned targets. While these programs add significantly to the increased capability of our missile systems to adequately defend their users, the expenditure of such sums is primarily for technological advancement of future systems, and do not necessarily lead to production orders for the industry involved.

Such was the case with the Firebolt and other target systems developed by TRA and other prime target producers. The case of expendable vs. recoverable targets promises to create continued competition within the industry for years to come.

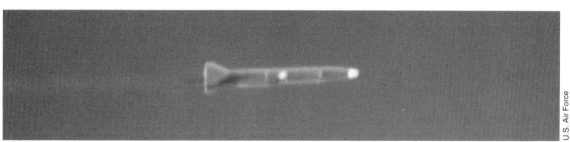

LAUNCHED ON ITS FIRST FLIGHT from an A4 Phantom *at 50,000 feet and Mach 1.5, the Firebolt later set a world record for unmanned sustained flight — speed Mach 4.3; altitude 103,000 feet.*

THE VENERABLE MARS Mid-Air Retrieval System *returns a hot bird to its land-based nest.*

SLAT

SUPERSONIC LOW ALTITUDE TARGET

Composite Photo

REQUIREMENTS IN THE '80s from both the Navy and Air Force were to fly higher — and faster; fly lower — and faster; and stretch the performance envelope further than at any time in previous history.

Competition within the industry was fierce. The leading proponents of unmanned flight spent thousands of man-hours in proposal presentations, in-house engineering design efforts, and customer briefings. Teledyne Ryan, with years of proven capability was a forerunner in several areas.

The supersonic Firebee II was an early leader in the high speed areas. And, the Navy's contract award for the Firebrand development program in 1977 provided valuable information on ramjet engines and improved booster bottles.

While no production orders followed the Firebrand engineering effort, the know-how gained put TRA into an advantageous position for the Air Force requirement for a high-altitude supersonic bird, and the later Firebolt system eventually topped 103,000-feet and hit Mach 4.3.

As to low-altitude, the ability to skim over the wavetops had been demonstrated many times to both services with the Firebee II and more recently the BQM-34S sea-skimmer modification program involving flights at 10 feet above 8-foot ocean ground swells.

When the Navy came out with still another low-altitude supersonic requirement in 1984, TRA was loaded and ready. Nicknamed SLAT (Supersonic Low Altitude Target), with **YAQM-127A** military nomenclature, the request for proposal arrived at an opportune time.

SLAT'S PRIMARY MISSION was to carry sophisticated threat replication electronics at Mach 2.5 at an altitude of 30 feet from a below-the-horizon launch point over 50 miles out.

With both Firebrand and Firebolt programs in abeyance, Vice President Richard E. Smith and 'Supersonic' Sam Sevelson were turned loose to head up the **Model 320** SLAT proposal team.

Reaching beyond TRA's specialties, United Technologies' Chemical Systems Division and the Defense Division of Brunswick Corporation were chosen as partners in the proposal effort. These team members had performed the same development tasks on the Mach 4 AQM-81 Firebolt program.

The Chemical Systems Division was given the task of providing ramjet propulsion with a solid propellant booster integral to the ramjet combustion chamber. Described as a swirl/spill-dump ramjet, its development and testing was funded jointly by TRA and United Technologies. With CSD's experience, coupled with TRA having integrated ramjet engines in the ZBQM-111A target system, the risk element in combining the engine and airframe in SLAT was considered minimal.

Brunswick had the responsibility for the ground test equipment, radome manufacture, recovery parachute module and launch aircraft interface. Design included interface with Navy F-4, A-6 and F-18 aircraft, and the DC-130 and P-3 launch planes.

Physically, the bird was not very large, with an overall length of 11 feet, a 17-inch diameter fuselage and small dual-configured elevators/rudders mounted aft on the fuselage. Weight, fully loaded, was less than 1,000 pounds. As a simulation for an enemy anti-ship missile, SLAT was designed to intimidate the U.S. Navy's latest ships.

Dick Smith was optimistic in his analysis of TRA competitive position. "I think we have an advantage in many areas as TRA has more experience in the integration of target system. SLAT is virtually a 'flying antenna farm' with a requirement to integrate a dozen electronic systems with 26 separate antennas into a subscale high temperature aircraft. Another plus is that we have the most supersonic low-altitude operational experience."

WELL — WIN SOME, LOSE SOME. Awaiting decision of the customer's procurement branch is always suspenseful and agonizing. When the award went to a competitor, TRA was shocked. Norm Sakamoto later recalled, "The award was the result of a competitive procurement process and we simply lost out.

"However, we submitted a proposal with a configuration that would take existing technology, and with some modifications to the propulsion system, would have been able to perform not only the SLAT missions but also 80 percent of the Firebolt mission."

And, once again, all the engineering design and system analysis had to be stored in the 'futures' vault to be used again. After all, TRA's approach to new contracts has always been to make maximum use of what has already been learned on every target system program.

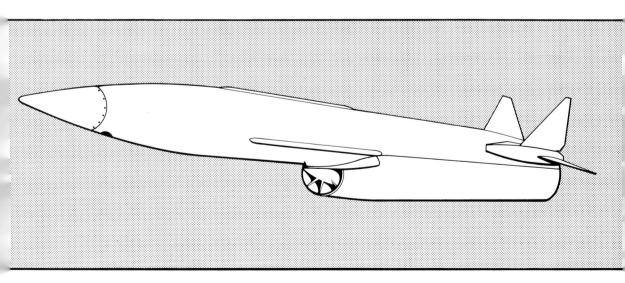

A flock of
NEW BIRDS

Artists rendering of High Altitude Long Endurance (HALE) 'Spirit'

*E*MPHASIS ON THE NEED *to develop new families of unmanned vehicles came from the military services during the 70s and 80s. Envisioned were a variety of reconnaissance, targeting and surveillance missions that could best be met with new hardware.*

Among the categories being studied: (1) mini-RPVs, (2) medium-sized or midi-RPVs, and (3) large or maxi-RPVs, the latter for high-altitude, long-endurance missions.

Each service had a different concept of requirements to be met. This, of course, led to some redundancy but that was before all procurement of Unmanned Aerial Vehicles (UAVs) was coordinated after 1988 through the Joint Project Office (JPO).

HUDSON B. DRAKE came aboard as Teledyne Ryan Aeronautical President in April 1984. He was not long in directing the company's research and development effort toward new unmanned vehicle projects. Already he had laid the ground work for resumption of production of Firebee target systems.

In a near unprecedented 'exclusive' to Aviation Week and Space Technology, Drake in spring 1985 discussed at length three areas of particular interest:

- High-altitude, long endurance RPVs (like Compass Cope) capable of serving as platforms for relaying communications, intelligence and target information.

- RPVs designed for tactical penetration missions.

- Mini-RPVs and other drones for use as targets and in surveillance operations.

As is well known, not all ideas generated in the aerospace industry result in contracts. However, new concepts and new design approaches, though not initially successful, in many instances lead to technological breakthroughs in later years.

TRA was indeed breaking new design ground with "A Flock of NEW BIRDS," but few led to production. Some, in fact, never flew and to that extent, were kiwi 'paper planes.' Others later led to major contracts such as for **Models 324** and **350**.

MID-RANGE PREVIEW

THE NAVY'S INTEREST in the so-called midi-RPV *(short for mid-range RPV)* was highlighted in a trade industry article in 1974 which outlined what the Navy considered to be its future requirements.

A decade later the midi-RPV was still on the Navy's wish list. When a Navy RFI — informal request for information — reached TRA, the company was well prepared to respond. Two years earlier a new design concept had begun working its way through company channels.

The illustration for an August 1985 article in TRA's employee publication provided the first glimpse of a basic configuration which would take shape in new production RPVs for a friendly foreign government and for the U.S. military services.

(See chapters starting pages 157 and 175.)

NEW CONFIGURATION for mid-range missions is seen for the first time in this photo of a model RPV. *At left, President Hudson Drake; right, Tom Brennan of Naval Air Development Center, leaving after a year's assignment in industry at TRA under an Executive Exchange program sponsored by the White House.*

'SPIRIT'

TEN YEARS AFTER completion of flight testing of Compass Cope-R, with its high altitude, long range capabilities proven and finally shelved by the Air Force, TRA was ready to try again in 1985 with a high flying drone tentatively named 'Spirit' and classified as a HALE (High Altitude Long Endurance) system.

The research, engineering and planning effort that went into the proposed **Model 329** Spirit was one of the most extensive company-sponsored programs ever undertaken. It was designed to meet several potential missions including communications relay, electronic intelligence gathering, sonobuoy monitoring for fleet anti-submarine warfare and long-range weather monitoring. Growth versions envisioned included over-the-horizon targeting and cruise missile early warning and tracking.

The Spirit was designed to have an 85-foot wingspan and an aspect ratio of 30. The fuselage would be 40 feet long, depending upon the type of payload carried in the modular nose, and the tail was 14 feet high. With its trim fuselage and graceful twin booms supporting an inverted 'V' shaped rudder, the bird was pleasing to the eye.

Designed to cruise at 50,000 feet with a 300-lb. payload and remain on station for 80 hours, its structure was to be almost entirely of composite materials. The prototype would be fabricated under subcontract by Scaled Composites, Inc. of Mojave, California. Load-carrying structures were to be made of graphite epoxy, with the wing leading section of Du Pont Kevlar and secondary structural components of glass fiber. Gross weight was expected to be 4,500 lbs., and empty weight around 2,500 lbs.

According to W.A. Grenard, senior vice president of engineering and development programs, initial flights would be manned with the pilot in a cockpit where the modular payloads would later be mounted in the nose section for normal drone operation.

The twin booms which support the control surfaces would also house the retractable main gear of the tricycle gear arrangement.

PROPULSION FOR SPIRIT was expected to be the six-cylinder 155-160 HP. Teledyne Continental liquid-cooled, turbocharged piston engine driving a 20-foot propeller mounted in a pusher configuration at the rear of the fuselage.

In a four-cylinder version this was the same basic engine that, rear-mounted, helped propel the Rutan/Yeager 'Voyager' on its history-making nine-day unrefueled circle of the globe in December 1986. [The Voyager's Continental air-cooled front engine was used mostly for takeoff, climb and landing as well as intermittently during the long flight.]

Evaluations were made for several navigational packages for long-endurance Spirit missions. Systems under study included an internal navigation system updated by the Navstar Global Positioning System satellites and an in-house avionics suite developed by Teledyne Ryan.

The talented engineering dreamers were in Shangri La! However, none of the military services were interested enough to pay for the prototype development.

Most interest in the concept was shown by the Naval Air Development Center but it could not find development funds to underwrite the venture.

A principal reason for lack of interest on the part of the military services was that there was not an off-the-shelf mission payload available. Thus, significant additional funding would be needed to justify developing a payload package to integrate into the Spirit vehicle.

Subsequently, the company embarked on Model 410, a little brother of the Spirit, born three years later. It was designed to carry an equal 300 pound payload but at lower altitude, with less endurance capability, and at a much lower procurement and operational cost.

WITH ITS 85-FOOT WINGSPAN and fuel efficient piston engine, *the 'Spirit' RPV would have high altitude, long endurance capability. But the cost of also developing a suitable payload package brought the project to a halt.*

THE MINI-DRONES

ALTHOUGH UNMANNED AERIAL vehicles had grown to large physical dimensions (81-foot wingspan of Compass Cope, for example), the whole idea grew out of small radio-controlled model airplanes. Film star Reginald Denny, a model plane hobbyist of the 1930s, was one of the mini-RPV's midwives.

Flying a model plane above the battle field, or at sea over-the-horizon, to see what the enemy was up to was an idea whose time had come.

Over the years, this led to many studies and development of a wide variety of airframe and engine mini-drone configurations for Army, Navy, Marine Corps and Air Force missions.

ONE EARLY VENTURE was Ryan's kite-like Flexbee, tested in 1962 under Marine Corps sponsorship. [Co-author Doc Sloan, a kite-flying enthusiast, managed the cesarean-section delivery of Flexbee.]

In Israel, where full-scale modified Ryan 124I Firebees were operated as RPVs during the Yom Kippur War of 1973, strong interest was created in developing and producing Israeli mini-drones of their own design. In this effort, Ryan technicians already on site offered a bit of friendly coaching now and then.

BY MARCH 1975, the Director of Defense Research and Engineering (DDR&E), reporting to the U.S. Senate, was calling attention to "mini-RPVs to be used for battlefield surveillance and target acquisition and eventually for laser designation of moving tank-sized targets for the employment of guided weapons."

Additionally, the report said, "The Navy is emphasizing shipboard launch and retrieval of mini-RPVs and their employment for surveillance and as airborne jammers."

MODEL 262 'MANTA RAY'

AN EARLY NAVY EFFORT was the STAR project, a mini-RPV demonstration program, supported by the Defense Advanced Research Projects Agency (DARPA).

Preceding award of a Ship Tactical Airborne RPV contract, Teledyne Ryan had flown experimental models RPV-004 and -007 as half-scale delta-wing testbeds. Using rotorduct propulsion, these drones flew conventional takeoff and landings during demonstration flights.

The final configuration which evolved from earlier half-scale tests, was **Model 262 'Manta Ray'**, of which three evaluation units were built.

With 7½-foot wing span and 25 hp piston engine driving the ducted propeller, the 160-pound delta-wing drone was easily transported by two men.

Constructed of fiberglass, and with virtually no straight lines or flat surfaces, the mini-drone had inherently low radar cross-section shaping and low infrared signatures,

COMPOSITE MATERIALS, delta-wing design and a rotorduct propulsion system were key features of the Model 262 'Manta Ray' mini-drone.

Bud Wolford

SHIP TACTICAL AIRBORNE RPV, known as 'STAR' *by the Navy and 'Manta Ray' by TRA, displays the sleek delta-shaped design which minimizes radar detection. Photo at right shows automatic net recovery 'landing.'*

assuring maximum survivability in hostile environments.

Designed for shipboard launch, the 262 utilized an All American, Inc., compressed-air rail-type launcher and was flown into a raised webbing for recovery.

A Poise optical tracker looking through the webbing provided automatic uplink commands to steer the air vehicle into the net. Unlike the hobbyist who has to wield a net to catch a butterfly, the Manta Ray had to be convinced it should fly *into* the retrieval web.

With six consecutive successful shore-based test flights, ending in net 'landings' at the Naval Parachute Test Range, the system proved the potential for operation aboard ship.

Other versions of the Manta Ray could, of course, be configured with conventional landing gear for takeoffs and landings.

This flight test program was successfully completed in 1976 at the El Centro, California parachute test facility.

MODEL 328 for ADAS

TRA TECHNICIANS FLIGHT TESTED Air Force's *Flight Dynamics Lab BQM-106 mini-drones in Utah during 1977.*

AN IMPORTANT AIR FORCE requirement is to have an available Airfield Damage Assessment System (ADAS). How better to provide it than by a developed mini-drone system? That's where TRA's **Model 328** came into the picture in 1985.

In the late 1970's, TRA built two of its version of the Air Force Flight Dynamics Laboratory (FDL) BQM-106. These mini-drones were demonstrated at Wendover Air Force Base in Utah under the direction of TRA's on-site test director, Billy Sved, who also handled the remote control tasks.

The need for an ADAS vehicle revived in 1985 with a new Air Force contract under which the company had Digital Design and Manufacturing hand build two vehicles to TRA design specifications, which evolved from the FDL BQM-106 drone.

To the unpracticed eye, the 328 with its high wing and high tail looked somewhat sim-

Dave Gossett

DESIGNED AS AN AIRFIELD DAMAGE ASSESSMENT SYSTEM, the high-wing high-tailed, propeller-pushed Model 328 is shown during flight test.

ilar to the post-WW II Republic Seabee personal plane design, except for its diminutive size.

Wing span was 12 feet; length 10 feet. Power was provided by a 25 hp. twin-cylinder engine and pusher-prop arrangement. Take-off of the demonstration vehicles was by a pneumatic-powered rail launch system. As an alternate, a zero-length rocket booster can be mounted beneath the tail.

The bird was also designed to use conventional wheeled gear for runway operation. A 24-foot parachute was provided for recovery on conclusion of its mission.

Included in the feasibility demonstration flights, with TV camera as payload, were those for a friendly foreign military service.

EQUIPPED AS A DATA gathering sys-

tem, the drone could house an infrared line scanner in its nose section. It could also be instrumented for airfield damage assessment missions for rapid collection of data on runway hazards and post-attack damage.

Typically, a real-time television surveillance mission would include launch from unprepared sites and programmed automatic flight to projected target areas. A remote control operator could take discretionary control via radio frequency command link when desired to investigate targets of interest or redirect the vehicle to alternate targets.

Depending on the mission and equipment carried, Model 328 has an endurance of over seven hours. Typical mission altitudes are from 3,000 to 10,000 feet with an operation ceiling of 16,500 feet. Airspeeds range from 80 mph at cruise to a dash speed of 120 mph.

POISED ON ITS PNEUMATIC-POWERED RAIL, the ADAS vehicle is ready for launch. Model 328 was designed for missions as long as seven hours.

Dave Gossett

Billy Sved

REMOTELY-PILOTED HELICOPTER

ARMY AND MARINE CORPS ground troops require training and experience in the use of shoulder-fired anti-aircraft missiles.

"In the case of the Marine Corps," reasoned TRA's Bill Grenard, "it does not have the long-range requirements of the other services, nor room to launch, recover and control vehicles.

"Why not use a remotely piloted mini-helicopter (RPH) for the missile firing training

ONE-QUARTER SCALE MODEL of Soviet Mi-24 Hind D *assault helicopter was flown in tethered tests to verify design objectives.*

Dave Gossett

Dave Gossett

mission."

Such reasoning led in 1985 to development of a one-quarter scale realistic target that accurately represented a Soviet Mi-24 Hind D assault helicopter.

THE HIND D VERSION of the RPH had a rotor diameter of 14 feet, fuselage length of 15 feet, and 285 pound gross weight. Production versions would make extensive use of glass fiber and other composite materials for the airframe.

Designed to carry 50 to 100-pound payloads, reach dash speeds of 55 mph and fly for up to five hours, the **Model 326** mini-RPH would have a range of up to 200 nautical miles. Service ceiling would be 10,000 feet.

Power for the one-quarter scale helo was a 50-hp ultralight aircraft engine.

Hover flight tests conducted in summer 1985 at Brown Field, San Diego, demonstrated very stable qualities according to program manager Gerald R. Kellar.

Engineering tests, including both tethered and untethered flights, verified design objectives. Lateral, forward and backward maneuvering in the hover mode were successfully demonstrated. For these demonstrations, designer John Simone was test pilot at the remote flight controls.

The capability was there, but how to find a customer among the military services able to provide contract support for production of RPHs?

TARGETS FOR LASER WEAPONS

LASER BEAM 'HITS' MODIFIED FIREBEE TARGET, "providing the most real-istic *test to date of the potential damage that can be inflicted on a missile by a laser system."*

U.S. Navy

IN THE HIGH-TECHNOLOGY era of 'Star Wars' studies and planning, what role might laser beams play?

Could a ship-based laser weapon destroy an incoming low-level cruise missile? What about a land-based laser intercepting and 'killing' a high-altitude enemy jet? What effect does the atmosphere have on beam propagation? And, could a laser be 'fired' effectively against an intercontinental ballistic missile?

A key installation in finding answers to some of these questions is the Mid-Infrared Advanced Chemical Laser (MIRACL) located at the Army's White Sands Missile Range (WSMR) in New Mexico.

For nearly four decades Teledyne Ryan Aeronautical has provided unmanned target aircraft and technical services at the WSMR installation. It was logical that TRA's unmanned jet vehicles should also have a role in laser research being conducted there.

SEALITE WAS A U.S. NAVY program of the early '80s which became the Department of Defense's major thrust to determine the feasibility of using high-energy lasers aboard ships for point defense.

The need for better defense against low-flying anti-ship cruise missiles was later confirmed, in May 1987, when the USS 'Stark' was hit by Exocet missiles launched against it in the Persian Gulf by Iraqi fighters during the 1982-88 Iran-Iraq War.

Key to high power laser weaponry was the MIRACL laser developed by TRW, Inc., under Navy contract. Static tests at TRW's San Juan Capistrano, California facility were conducted against Navy-furnished, instrumented BQM-34S Firebees to determine whether and how laser beam weapons might eventually be installed aboard ships to provide better defense.

The MIRACL laser equipment was moved in 1983 to the High Energy Laser System Test Facility at White Sands. There the laser beam generator was mated to Hughes Aircraft's precision optical beam director for realistic anti-ship missile defense tests.

Because of the basic capabilities of the BQM-34S Firebee and, incidentally, long-standing operational experience at WSMR, TRA was selected, along with other drone makers, to provide targets capable of simulating hostile missile attacks for the laser weapons tests.

139

Under contract to the Naval Air Systems Command, TRA customized a prototype **Model 314** version of the Navy BQM-34S, and built additional production vehicles based on the prototype.

Externally, the special purpose Model 314 was just another training target to be expended when 'killed' during testing of a new weapons system.

Internally, the 314 was loaded with sensor arrays at designated 'aim points' and other specialized instrumentation to record effects of the laser. The beam director would track the target and point the laser beam at the desired aim points.

Instrumentation down-link systems transmitted pertinent flight test data to ground technicians until final impact, or mid-air explosion, of the Firebee as its in-flight systems became inoperable.

In addition to providing analysis of the effects of laser beams on critical aircraft components and subsystems – and flight dynamics of the vehicle – the tests made it possible to also explore problems of propagating high-intensity laser beams through the atmosphere.

THE FIRST OPERATIONAL test, September 18, 1987 was designed to destroy the target flying an offset pattern at a distance and with the laser power that would demonstrate the military application of the system in a tactical environment.

After ground launch, while operating under its RALACS radar altimeter low-altitude control system, at an altitude of 1,500 feet above ground level, the target was intercepted and destroyed by the MIRACL laser in the lethality test.

It was, the Navy said, "the most realistic test to date of the potential damage that can be inflicted on a missile by a laser system."

In a November 2nd test, the second Model 314's vital components were similarly destroyed by the laser at twice the range of the first flight test.

According to the aviation trade press, by spring 1988 five miniature 'killed' Firebees had been painted on the Hughes beam director equipment at White Sands.

"During tests of the 314," reports B.R. (Bob) Hamrick, TRA engineering program manager, "the instrumentation performed flawlessly in providing needed data from the 'aim points' installed in the targets."

THE 336 HATS BIRD

EVEN AS LASER TEST work was getting under way at WSMR, the Navy and ever-changing government agencies involved in Star Wars and Strategic Defense Initiative planning saw the need for further advanced testing.

More should be investigated, these agencies reasoned, about atmospheric propagation of high energy laser beams and their possible future application against intruding weapon systems approaching at high altitudes.

And, this time couldn't TRA provide an airborne platform carrying a large sensor array to transmit data to ground stations without having the target destroyed by the laser beam?

SPACE AND NAVAL WARFARE Systems Command contracted with TRA in March 1986 for a 30-month program involving a new **Model 336** vehicle which would provide a stable flight platform, configured with sensor arrays and able to operate at 50,000-foot altitude.

This program had the code name High Altitude Target Skylite (HATS).

Again, a BQM-34S Navy-type Firebee was modified for the test work, with additional vehicles to follow development of the prototype.

Configurations considered were a ping-pong-table size array suspended beneath the drone fuselage. But first, data from wind tunnel tests showed that a better aerodynamic approach was two 53-inch diameter discs, one suspended beneath each wing by two short struts.

Over 300 sensors were provided in the 31-inch diameter active sensor arrays. These 2½-inch thick sandwich arrays included the ½-inch thick gold-coated aluminum exterior

SENSOR ARRAY MOUNTED UNDER THE FIREBEE'S left wing *is designed to reflect most incident energy from a ground-based laser.*

Dave Gossett

DESIGNED TO SURVIVE TESTS of ground-based MIRACL laser, *Model 336 target vehicle carried two 31-inch sensor disc arrays for high altitude tests.*

Dave Gossett

panel, as well as a gold-coated copper aperture panel and other supporting layers.

Initial plans called for a high-energy sensor disc beneath the left wing, and a low-energy sensor under the right wing, but budget considerations eliminated the low-power unit. As a result, a dummy disc was provided under the right wing to assure aerodynamic symmetry in flight.

Sensors behind the conical apertures of the array, which are the aim point of the beam, sample the laser's thermal profiles. This data is processed by on-board instruments and telemetered to the ground station.

THE HIGH-POWER ARRAY, with its gold-coated surface pointing downward, is designed to reflect most of the laser energy. Only a small amount reaches the thermocouples behind the protective reflecting surfaces. To assure safe return of the target, its fuselage is unpainted and highly polished to reflect the otherwise damaging laser beam.

After the beam locks onto the high-altitude drone underwing sensor array, the ground-based MIRACL laser is fired for only a few seconds.

A proving flight to 53,000 feet was made July 23, 1990 at White Sands to validate operational aspects of the ground-launched Model 336 prototype vehicle. The 336 ex-

ceeded all specifications and the program appeared to be on its way.

Although the laser weaponry program has seen periodic changes in its sponsoring agencies and diminished funding, many of those originally working in this field have continued to be involved as the organizational structure has of necessity been altered.

Currently the newly configured laser test birds are in a flight-ready condition awaiting the next step in the continuing development of the laser systems.

"We're at the beginning of the process," says Hamrick. "As we learn, the targets and sensors will become more refined and more practical. We, and the Navy, have the benefit of what research and development has already been accomplished."

HATS HIGH ALTITUDE TARGET SYSTEM'S fuselage *was highly polished and unpainted to reflect otherwise damaging laser beam.*

U.S. Army

VIETNAM-ERA AQM-34L ready to fly, *and better than new!*

Dave Gossett

Out of the Warehouse and
BACK TO WORK

O*NE OF THE MOST DISMAL after-effects of any war is the destruction of remaining, unused inventories of aircraft. Following World War II, thousands of aircraft were either 'mothballed' for limited periods, sold or destroyed.*

At the conclusion of the Vietnam conflict, a substantial number of the reconnaissance RPVs from bases at Bien Hoa, U-Tapao in Thailand and Osan, South Korea were shipped home and warehoused, mainly at Warner-Robins AFB in Georgia and at Hill AFB in Utah.

What follows is an account of how 53 war-weary birds restored in 1987 were revived for useful new tasks after a dozen years in the warehouse.

Three tasks have been undertaken by the Air Force's 6514th Test Squadron in support of:

- *Over-the-Horizon Radar*
- *North Warning Radar System*
- *'Aegis' Cruiser Tests*

OVER-THE-HORIZON

THE STORY OF REEMERGENCE of the Vietnam RPVs into an active role with the military services begins at the Air Force Systems Command at Hanscom AFB in Maine. The basic mission of this command is to perfect radar detection systems to warn the Continental U.S. of any incoming hostile aircraft or missiles.

Authorization to improve on the obsolete radars of the late fifties and sixties was given to the command in early 1972. The new system was to be known as Over-the-Horizon Backscatter (OTH-B).

Conventional radar operates by broadcasting radio beams, then listens for echoes when signals bounce back from distant objects. The range of radar is limited by the earth's curvature, as the signal follows a straight line. Ground-based equipment has an operating range of about 50 miles and airborne systems 200 miles.

Over-the-horizon radar by contrast sends beams up to the ionosphere, an atmospheric layer of charged atoms that begins some 45 miles above the earth's surface. The signals are then reflected by the ionosphere over the horizon where they hit objects and bounce back to the ionosphere, then back to the radar system.

The idea behind OTH radar is not new; shortwave radio signals have long been bounced off the ionosphere. Developing reliable over-the-horizon radar was more complicated because the composition of the ionosphere is always in flux but this problem was overcome by using computers which enabled operators to determine where conditions are best, and which frequencies could be used for best performance.

Built by General Electric, the system can detect targets up to 2,000 miles away, nearly ten times as far as conventional radars. To be effective, the OTH radar system is huge. The signals are sent from three transmitting antennas, each more than 3,600 feet long, in Moscow, Maine. Some 110 miles away in Columbia Falls are three receiving antennas each stretching nearly a mile. Twenty-eight computers located in the operations center at Bangor, Maine control the whole system.

Initial tests of the OTH system in 1986 revealed that while high-flying aircraft, including commercial airline traffic, could be easily detected, cruise missiles or low-flying hostile aircraft were extremely difficult to detect.

A POSSIBLE ANSWER to testing the ability to detect low-flying cruise missiles or jet fighters would be to fly RPVs, modified to have reflectivity characteristics similar to Soviet AS-15 cruise missiles, in low-level simulated attacks. Thus a request for bids to rehab and modify existing targets in storage was sent out in October 1986 to five potentially qualified companies.

Teledyne Ryan Aeronautical was able to make a prompt response since the statement of work clearly indicated its ability to perform. TRA's field service people were already available at Hill AFB in Utah; they knew the condition of the warehoused assets and knew how to rehab them for new missions in the shortest time and at the least expense.

Utilizing current inventories with minor 'quick reaction' modifications, instead of developing new equipment, had a strong appeal to the customer. With experienced civilian talent available to start the program, military personnel could be trained for eventual takeover for expanded operations. This assured the Air Force that it would not be solely dependent upon a contractor-run system.

And, with experience, the 6514th Test Squadron at Hill would be able to furnish similar services to other branches of the military.

Rip Van Recce awoke to a new world!

A CONTRACT TO REHAB, test and modify 18 of the vintage **Model 147SC** and **SD** birds soon followed. Hill AFB was chosen as the location for the rehab and test programs. Veteran Billy Sved was named TRA's Project Manager.

With preliminary inspection, replacement of a few storage-weary units and the addition of several state-of-the-art electronic modules, Billy stated "They're ready to fly, and better than new!"

Surrounding Sved was the usual 'can do' cadre of contractor field reps, veterans of the Vietnam era, and 'new kids' with hundreds of hours of experience in maintaining, control-

ling and engineering RPVs. Welcoming them at Hill were also a group of Air Force pros, and the old rapport and camaraderie between RPV addicts was quickly formed. An embryonic nucleus of future programs began to take shape.

The flight test program included air-launched RPV missions ranging up to one thousand miles over the Utah Test and Training Range, with altitudes varying from 500 feet to over 40,000 feet. The birds not only repeated the successful missions flown a dozen years or more earlier in Vietnam, but did them better.

The successful conclusion of the test program out of Hill AFB found the system ready for Phase II evaluation in over-water flight out of the Navy facility at Roosevelt Roads, Puerto Rico.

TRANSFER OF MEN AND EQUIPMENT from Hill AFB to Rosie Roads was smoothly accomplished, using NC-130 Hercules transports. Ground control equipment was housed in refurbished and updated vans capable of fitting into the belly of the big aircraft. The 147 model RPVs (military designations **AQM-34L** and **M**) were either carried externally on racks or stored inside. Billy Sved recalls, "We moved everything but the hangars."

The distance between San Juan, Puerto Rico and the island of Bermuda is over 1,200 miles. At altitudes ranging from 500 feet to 25,000 feet, the RPVs masquerading as hostile cruise missiles would be flying various trajectories toward the Bangor radar.

To simulate hostile aircraft or cruise missiles at extreme ranges required a ground control station located atop a peak on St. Thomas in the Virgin Islands, a little east of San Juan. This station took over from the airborne control station aboard the C-130 launch aircraft and directed the bird to a precise landing in

THE CAN-DO CREW OF THE 6514TH Test Squadron prepares to leave Hill AFB in the C-130 Hercules for assignment at Roosevelt Roads, Puerto Rico.

Dave Gossett

the recovery area to be picked up by a Navy chopper and returned for decontamination and turn-around.

The C-130s provided airborne control over the Bermuda area. Refurbished birds were then flown back to Rosie Roads on the C-130 for complete turn-around. With two ground control vans and remote control from the operators aboard the launch aircraft, the entire flight was under full control at all times.

SEVERAL PROBLEMS EXISTED in operating hundreds of miles from land. Frank Oldfield, TRA's Director of Engineering, recalls, "We felt we might lose control of the birds over the open ocean, so we put a piece of equipment in that would fly them on their own for five or ten minutes. If they descended below 500 feet, they would automatically climb to 10,000. If command was lost beyond a pre-set time, the chutes would automatically pop out."

Since the Navy did not have any helicopters available or configured for the standard Mid Air Retrieval System, the decision was made to water-recover the birds at the completion of their missions. An agreement was made by TRA with NASA for use of recovery ships utilized by NASA in case of early failure of a shuttle flight launched from Florida. Contractor personnel aboard the boats performed preliminary decontamination after pickup. Post-recovery locator beacons guided the boats to the drone, but distances and weather sometimes thwarted retrieval of the drones.

Oldfield continued, "Out of 32 launches from the C-130s, we had 29 successful flights during that six month period. Six birds failed to be recovered either because weather prohibited the NASA boats from going out to sea, or because of distances beyond a reasonable range of the craft."

The survivability of the targets in the open seas had been demonstrated many times in the past. Navy Firebee training targets downed in the ocean off Pt. Mugu in Southern California have floated for months, with intact equipment compartments, as far away as Hawaii and beyond.

Frank related the saga of one 147 bird which disappeared in the Bermuda Triangle. It was picked up by a Spanish freighter which eventually arrived in New Orleans. Attached to the drone was a message reading "If found, please call this name and telephone number. There may be a reward."

Fortunately, or otherwise, Oldfield's name and phone number were the ones listed. Through a series of phone calls by the Spanish skipper, the Air Force at Tyndall AFB in Florida was contacted, and the Air Force said

DECALS ON FUSELAGE OF AQM-34M RPV represent more than 30 *successful parachute recoveries, most of them earned while operational in Vietnam.*

"thanks, but no thanks — you can have the darn thing." Undaunted, the skipper kept it aboard through the Panama Canal and eventually dumped it on a Spanish frigate. Contact by the Spanish officials to Air Force officials in the U.S. weren't very productive until the International Maritime Rules were evoked, giving the salvage country payment for costs incurred. A token settlement for around fifteen thousand dollars was finally arranged.

THE TEST AND EVALUATION of the TRA 147s (AQM-34Ms) at Roosevelt Roads proved the capability of the RPVs and personnel involved to operate at great distances over the open ocean. At the completion of 32 flights, the Air Force sent letters of commendation to Teledyne Ryan expressing complete satisfaction for the success of their mission.

The stage was now set to fully integrate the actual test of incoming low-level targets vs. radars, and the next move was to Maine for actual hands-on operation.

November 17, 1988 was a milestone in the history of the rejuvenated 147 drone program. An AQM-34M was launched from an Air Force NC-130 over the ocean off the coast of Maine and flew a successful mission, testing two of the government's prime defense systems.

This flight was the culmination of years of preparation by the Air Force, Navy and Teledyne Ryan. The Navy was the primary customer for the 6415th RPV services as 'Aegis' class cruisers launched from shipyards in Maine needed to test their missile defense equipment during initial sea trials.

Concurrently, the Air Force used the flights to test OTH-B radar's short-range effectiveness over the North Atlantic.

Four different branches of the services worked together in harmony, each contributing expertise to assure that their responsibilities would be met without flaw. The Air Force supplied the launch aircraft, manned by both TRA and AF personnel, while their radar technicians operated the OTH-B systems near Bangor.

The Navy supplied the recently launched missile cruiser 'Philippine Sea' and escorts to test the Aegis system. Maine's Air National Guard provided the maintenance facility for the drones and personnel to assist in preparing the targets for the flight. And the U.S. Coast Guard was on hand to recover the bird after its water landing.

The Philippine Sea fired two unarmed missiles at the target which had been launched earlier, at 0900 hours. Each missile scored a near-miss which would have been lethal had they been armed. Recovery was commanded, and the Coast Guard picked the bird out of the sea for return to Bangor.

The celebration at day's end was justly deserved by all participants. This single flight gave sufficient evidence that all segments of this test were ready for further deployment and operational use. The decision was made to regroup and look forward to uses of the Air

Dave Gossett

AIR NATIONAL GUARD AT BANGOR, MAINE provided maintenance *facility and personnel for Over-the-Horizon tests flown by the visiting Air Force drone squadron.*

Force-operated launch aircraft and its RPVs in different locations and environmental surroundings.

Participating in every phase of the operation, TRA personnel, under the guidance of Billy Sved as program manager, demonstrated again their special ability to work in complete understanding with their service colleagues.

'NORTH WARNING' ALASKAN TESTS

VALIDATING THE EFFECTIVENESS of northernmost U.S./Canadian early warning radar systems in their ability to locate and identify approaching enemy cruise missiles was yet another task for the 6514th-TRA group.

To replace aging Distance Early Warning (DEW) radars, a new North Warning System of 15 manned long-range and 39 unmanned short-range stations is being developed by the Air Force Electronic Systems Division, Hanscom AFB.

Dave Gossett

THE PROGRAM CODE-NAMED 'Gap Filler' explains the reasons for this requirement. It was learned shortly after activation of the new installation that low-altitude 'blind spots' existed in Long Range System coverage on the northern edges of Canada and Alaska looking toward the Arctic Circle.

Using some of Hill's RPVs was critical in assessing the Short Range System's ability to classify, track and provide intercept data for incoming threats.

The RPVs would masquerade as low-flying

NEW MANNED RADAR installations facing the Arctic Ocean *off Alaska are replacing older Distance Early Warning (DEW) radars.*

UTAH-BASED AIR FORCE NC-130H Hercules carrying two RPV *flies over radar installation at Barter Island. Fifteen missions were flown over a 22-day period.*

enemy cruise missiles and would operate at different speeds and angles relative to the radar station.

Because the Gap Filler radars are unmanned it was thought, during ice break-up and spring migratory bird activity, that the radar might give false alarms of low-flying enemy targets such as cruise missiles.

Since Alaska is known as the land of the midnight sun with summer daylight existing almost 24 hours daily, the period of May through July, 1989 was chosen to take full advantage of daylight and to test operation during the ice-breakup and heavy bird migrations.

Hill AFB was again selected as home base for the operation, but the move of the 6514th Test Squadron from there to Alaska was different from previous deployments. Rather than run several shuttle flights with the NC-130s, two giant Air Materiel Command C-5As were used to transfer the bulk of equipment from Hill to Elmendorf AFB, Anchorage, Alaska.

From there, the MAC C-130s and one of the 6514th cargo C-130s shuttled men and materiel to Barter Island, off the northern Alaskan coast facing the Beaufort Sea and Arctic Ocean. Distance covered from Hill to Barter Island was 2,500 nautical miles.

Despite nearly 24 hours of daylight, the remote Barter Island outpost, adjacent to the tiny village of Kaktovik, Alaska, would not be considered an ideal place for a summer vacation. High winds, frequent snow squalls, normal temperatures running between 30 to 38 degrees, and living quarters comparable to the Quonset huts of WW II. However, living conditions did not deter from the prime purpose of the tests.

Of the eight **AQM-34L** RPVs assigned for the operation, all eight returned to Hill at the completion of the tests although several small problems were encountered.

After hour-long flights well north of Barter Island over the Arctic Ocean, ground control would activate deployment of the RPV's recovery parachute. The drone would then drift to a soft landing on the Arctic tundra at a recovery site 40 miles west of Barter Island.

An H-3 helicopter from the 71st Air Rescue Squadron at Elmendorf would pick up the drones in a sling and carry them back to Barter Island to be recycled for more flights.

PYLON-MOUNTED AQM-34L will soon be airborne *and ready for air launch. Helicopter of 71st Air Rescue Squadron stands ready to pick up the RPV after its descent beneath recovery parachute.*

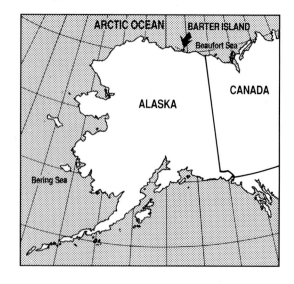

146

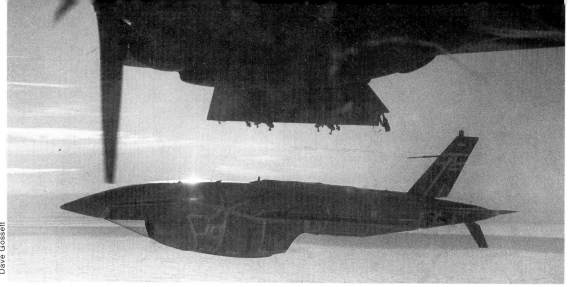

RETURNED TO DUTY 14 YEARS AFTER VIETNAM service flying 'spy' missions, *an AQM-34L is launched from Barter Island, Alaska to test how well new radar system can detect hostile aircraft flying at low and medium altitudes.*

Because of the quick turn-around times provided by the maintenance crews, the eight-week test program was completed a week earlier than scheduled.

Upon the RPVs landing on the snow pack or the tundra, the main parachute did not always detach upon contact because the release probe would sink into the tundra, rather than pop the chute loose. Minimum damage resulted, yet helicopter pick-ups by the Elmendorf choppers were successful in all recoveries.

Dave Gossett, the TRA photographer on hand at the pick-up site remarked, "One bird, with chute still attached, looked like it had floated downstream over the soft tundra for at least four miles — but there was hardly a scratch on it."

And, one day was lost when a C-130 support aircraft got stuck in the mud at the end of the runway!

PREPARATION FOR MISSIONS by the 6514th and the accompanying TRA technicians was thorough. All spares and equipment necessary to operate the birds were available as needed. On the other hand, small parts required for the maintenance of the C-130s and the helicopters occasionally caused program delays.

It wasn't a matter of running down to the corner auto parts store for a gasket. O-rings for a leaking oil fitting on the launch plane or tail-rotor replacement parts for the helos had to come from supply depots 2,500 miles away.

Despite minor mishaps, the entire program ran exceptionally smooth. Daily reports from Billy Sved to Gene Motter, his home-base counterpart in San Diego, tell of a typical mission on June 12th.

"Test objectives: Presentations to NWS Radar at 5K feet, 60 nautical miles for acquisition of additional detection and tracking accuracy data.

"Summary: Two successful AQM-34L missions against NWS radar installation at Barter Island, Alaska. Mission flown in snow storm. Both missions controlled by MCGS ground station; air director inoperative due to malfunctioning of the receiver/transmitter unit. Recoveries and retrievals satisfactory. First flight vehicle already recycled for another flight. Data acquisition, launch, communications and retrieval all satisfactory."

A total of 16 missions, all successful, were flown between May 13th and June 27th, 1989, and return of all personnel and equipment was accomplished in reverse order back to home base without incident.

A communication from Capt. Larry D. Wineteer, the Deployment Commander out of the 6514th Test Squadron summed up the performance of the operation.

"The support provided by the people at Barter Island was exceptional in all details. The excellent support from the 71st Rescue Squadron H-3s deployed out of Elmendorf AFB was the key in the on-time completion of this test.

"The individual performances given at all levels were at all times outstanding. The supervisors used lessons learned from previous deployments to great advantage. The ability to perform at such a high level in a very austere and demanding environment demonstrated an exceptional dedication in getting the job done and all should be commended for their efforts."

147

PRIOR TO AN AEGIS MISSILE FIRING TEST, a DC-130A of the Air Force *6514th Test Squadron, carrying two AQM-34s, passes over the USS 'Chosin' cruising in Hawaiian waters off the Pacific Missile Range Facility at Kauai.*

TARGETS FOR 'AEGIS' CRUISERS

AEGIS — *GREEK WORD meaning to protect.*

Aegis — a U.S. Navy surface-to-air radar defense system named for the shield of the Greek god Zeus.

Cruisers have a new importance in the Navy because of the installation of the Aegis missile tracking and guidance system which is installed on 'Ticonderoga' class cruisers. With cutting-edge technology, the Aegis system oversees the detection, tracking and elimination of as many as a hundred targets at a time in a battle zone.

ALTITUDE AND SPEED of a target, whether very high or very low, is of no consequence to Aegis. It can detect, track and kill a wide range of attackers.

The Aegis defense system got one of its first tests off Southern California July 13, 1987. Three **BQM-34S** training targets were air launched and sent aloft, flying in formation to mimic hostile invaders to be fired at by armed missiles directed by Aegis system radar.

The USS 'Bunker Hill,' preparing to leave for duty in the Middle East, was equipped with the vertically launched Standard missile. Captain Phil M. Quast, skipper of the Bunker Hill was favorably impressed with the performance of his ship killing the three Firebees.

"The first missile intercept," he said, "was made as the target, simulating an 'enemy'

cruise missile was at approximately 150 feet altitude, and occurred within eyesight by observers aboard our ship."

Later, the system was put to test during an actual 'combat' situation. In pre-Desert Storm days in the Persian Gulf during the Iraq-Iran War, the lethal capability of the cruiser-based Standard missile was demonstrated.

It was in the Gulf that the Aegis-equipped USS 'Vincennes' accidentally downed the Iran Air commercial airliner which it mistakenly thought to be an approaching hostile military aircraft. (Not long before that an Exocet missile launched by an Iraqi fighter had struck the USS 'Stark' on patrol in the Gulf.)

Indirectly, but noteworthy, the Firebees earlier played a role in preparing Aegis missile ships for action in the Libyan challenge in the Gulf of Sidra in April, 1986. The Navy Mobile Sea Range had exercised the USS 'America' and USS 'Ticonderoga' in a simulated Sidra area 500 miles north of Puerto Rico. Later, America and Ticonderoga battle groups joined up with the Sixth Fleet after a direct transit through the Straits of Gibraltar and into the Mediterranean Sea following the exercise.

A still later test of Aegis, November 17, 1988 against a rehabed **147SC** (AQM-34L) had been combined with an evaluation of Over-the-Horizon Backscatter radar off the coast of Maine, with take-off of the NC-130 launch aircraft from Bangor.

PRE-DAWN CHECK-OUT OF C-130 shows RPV drone slung under each wing of the launch plane.

NED PORTER, REPORTER for the Bangor News, was allowed to ride in the launch plane, and his excellent coverage of that flight portrays the crew's feelings that morning.

"From a left window," he wrote, "the day's first light creased the eastern horizon. On board the Air Force plane, crew members not flying or sleeping sipped coffee and waited.

"The C-130, a four-engine workhorse that the military has adapted for a multitude of uses, had taken off a couple of hours earlier under a clear night sky. A 27-foot drone was slung under each wing. The hold was a warren of electronics to launch, monitor and control the remotely piloted vehicles.

"It was heading south past the tip of Cape Cod for a rendezvous east of Long Island with the USS 'Philippine Sea,' a Navy Aegis cruiser undergoing sea trials.

"A contingent from the 6514th Test Squadron had flown from Hill Air Force Base in Utah in two C-130s for the mission. Much of the squadron's duties could not be discussed, Lt. Col. Wesley White, the deployment commander, said on a preflight tour of the aircraft.

"The mission flown from Bangor gave the Navy the chance to test the ability of the system aboard the Philippine Sea to counter sea-launched cruise missiles.

"'I've been launching these things for damn near 30 years,' Bill Sved, a civilian engineer with the drone's manufacturer, Teledyne Ryan, said while watching crewmen half his age.

"Moving through the plane with a familiar ease, Sved found some space, lay down, used a parachute for a pillow and nodded off until launch time.

"The crew had been prepared to launch the drone by 6:45 a.m. but the range was not secure. A freighter was still in the area and the crew on board the C-130 stood by for two hours for the ship to clear the range. While waiting to 'drop the bird,' people passed the time reading, snoozing, or sipping coffee and nibbling a doughnut.

"At 8:45 the word was passed: '15 minutes to launch.'

"The slumbering C-130 sprang to life. The civilians and crew members huddled around the four consoles from which the drone was controlled. The converted cargo bay became a hive of activity as people ran the equipment through one final test.

"The pilot, Capt. Larry Wineteer, banked left and right, and then climbed and dived in a parabola so the crewmen who would monitor its flight could ensure that the guidance equipment on the starboard drone worked.

"**E**IGHT MINUTES TO LAUNCH and the radio traffic — between ship and plane, between the crew flying the plane and crew flying the drone, among the men at the controls of the drone — continued increasing.

"At the command to launch, the **AQM-34L** drone dropped from the wing and disappeared from sight. The action in the C-130 intensified. The bird flies unmonitored until the radar officer locks it in, which took a couple of minutes. A dozen pairs of eyes were trained on the controls.

"Tensions eased when a small green blip appeared on the screen.

"At the control console, Capt. Robert Fitch took the drone through its paces using controls that resemble a sophisticated video game. 'I used to fly remote-control airplanes, which was kind of exciting, but nothing like this,' he said after the mission.

"Although the drone had trouble flying left turns, Fitch piloted it into range and the Aegis' system on the Philippine Sea launched a missile. With no direct hit, the drone was brought around for a second presentation.

"Orders came down to delay the second launch until a supersonic Concorde flew through the missile range.

"On the second run, the cruiser launched another missile and the bird remained alive on the control screens. An escort ship, the USS Clark got its turn and fired a shot.

"The command came to deploy the parachute for a splashdown.

"A small chuckle of competitive glee from the engineers, officers, and non-coms broke the tense atmosphere in the plane.

"'The preliminary word we got on the sea trial was excellent,' said Jim McGregor, spokesman for the shipyard that built the cruiser. 'We were two for two.'

"'Metal on metal contact' is not necessary for a good test, said a technician aboard the C-130.

"'Success is gauged on coming within a certain distance. They were hits,' McGregor said. 'The drone would have been destroyed if

ABOARD LAUNCH PLANE, TRA chief photographer Dave Gossett *catches action shot as the RPV is released from the wing pylon and quickly disappears from sight.*

Dave Gossett

the missiles were armed.'

"Flying through the test area before heading back to Bangor, Wineteer took the plane over the Philippine Sea and then over the Coast Guard cutter that had been standing by to recover the drone which was bobbing on the sea in the midst of its orange parachute.

"On the way back, Marvin Muklebust, a civilian engineer, walked forward with some cargo he had stashed. Setting the plastic box on some stacked cartons, he pried off the lid and offered up the 10 pounds of shrimp he had brought along to make the return flight a little more pleasant."

THE HARMONIOUS Navy-Aegis-Air Force-RPV relationship has continued since November, 1988, with exercises off-shore in the Hawaii area, Puerto Rico, and the Pacific Missile Test Range at Pt. Mugu. Additional training operations are planned by the Navy well into the next decade.

An additional C-130, refurbished throughout to improve the launch and control capability of the targets, joined the Air Force inventory in January, 1990 to ensure greater reliability in future exercises for Navy Aegis testing.

During 1990 the 6514th Test Squadron was busy living up to its slogan "Have drone, will travel". In January they again flew for the new Aegis warships at Bangor, Maine.

March saw them at Puerto Rico again, with

TECHNICIAN BILLY SVED explains for television *significance of Air Force mission supporting Aegis cruiser missile test firings off coast of Maine.*

Dave Gossett

five flights conducted with systems including low-level missions, ECM chaff dispensing and a new miss distance indicator. Once more the Navy expressed complete satisfaction with the Test Squadron's and TRA's technicians performance.

With only a single day's layover, the aerial caravan flew to Hawaii's Barking Sands range on Kauai Island. Two satisfactory flights were flown before a return to the Continental U.S., then down to Rosie Roads for an additional five flights. September saw them back at Bangor for two flights against newly commissioned ships with Aegis systems.

An official Navy "well done" signifies the important role that the AQM-34L/M and BQM-34S Firebees have played in the development, deployment and battle readiness of the Aegis defensive system. With the Persian Gulf crisis resulting from the Iraqi take-over of Kuwait, the Navy ships deployed to guard the Gulf were well prepared to perform their roles, either offensively or defensively.

As the Navy has written, "The planned Aegis Test & Evaluation (T&E) effort cannot be sustained without the Hill AFB RPV operations capability."

THE CLOSE INTER-SERVICE coordination between Navy and Air Force in scheduling Aegis and Over-the-Horizon (OTH) radar tests is noteworthy.

Both rehabilitated Air Force AQM-34L and M Vietnam-era RPVs, and more recently ordered Navy BQM-34S targets are being made flight-ready and DC-130A-compatible by the Hill AFB support team. A multiple over-the-horizon drone control system to support AQM-34L/M and BQM-34S operations with DC-130 launch plane capability is being provided. This will permit more complex open ocean and multi-ship exercises.

Some of the required Aegis T&E flights are to be conducted by the Navy with staging at the Pacific Missile Test Range facilities, Pt. Mugu.

NEW GENERATION UAVs

A milestone year in development of unmanned aerial vehicles,
1988 saw three new Teledyne Ryan RPVs in flight test:

TRA Model 410 - the 'affordable' RPV
TRA Model 324 - the 'Scarab' RPV for Egypt
TRA Model 350 - the Mid-Range UAV for the
U.S. Navy, Marines, and Air Force

Dave Gossett

Configured for unmanned flight, the updated Model 410 left the 'runway' at Cuddeback
dry lake, Edwards AFB, July 25, 1992 on its successful first flight and landing.

THE AFFORDABLE MODEL 410

PROSPECTS FOR A GLOWING future in the UAV business in 1985 were not too promising. While Teledyne Ryan was comfortably assured of continued orders for the 'old reliable' subsonic Firebees, future applications of other unmanned systems were overdue for investigation and exploration.

Hudson Drake, recently appointed President of the company, took a long range approach to the problem and initiated a two-year study to find a viable slot for, among other UAVs, an affordable vehicle that would fill a requirement for the military, other governmental agencies and potential civilian markets.

Project
Engineer
Doug
Fronius

Dave Gossett

AN EARLY APPROACH to an affordable RPV, especially for potential foreign governments, was the company's proposed Mulit-National Model 400 RPV.

The study program was centered around modifying an existing two-place canard-type piston engine aircraft, the Rutan 'Varieze.'

The planned modification would convert the home-built piloted aircraft to unmanned configuration. However, to make the proposed low-cost program successful, the existing design would require factory-rated tooling, manufacturing processes and quality control for production orders.

In addition, a different configuration than the Varieze appeared to offer more advantages as a UAV.

A wholly new start from scratch to accomplish the same end result seemed more realistic.

MODEL 410 was a company sponsored development to be based on worldwide research on the capabilities of RPVs and UAVs currently offered by many countries, and the needs and applications of potential users of such vehicles.

To direct the project, the company brought aboard a well-known industry executive with prior unmanned aerial vehicle experience. He and an assistant would travel widely and submit results of their interviews with knowledgeable sources both overseas and domestic.

"There's a great market out there," they concluded. "More than a dozen countries are already looking for truly affordable RPVs."

Designed as a multi-role vehicle powered by an off-the-shelf general aviation gasoline engine, the 410 would fill the gap between mini-RPVs and long endurance, high-flying 'spy' planes.

The first phase of their study revealed that unmanned vehicles were being used at only a small fraction of their potential. Existing systems, for example, provided a practical means

of conducting sustained surveillance but not at a really affordable cost. High operational costs limited continuous use of most sophisticated UAVs.

The second phase of research involved exhaustive analysis of over 100 types of missions for UAVs, coupled with studies of the tradeoffs required in vehicle design to accommodate the majority of these missions. This research was based on extensive interviews and discussions with potential customer audiences throughout the world.

What emerged from this study was a need for full-time, all-weather surveillance of borders, coastlines and vast land and sea areas.

Large payload space and weight carrying ability were seen as necessary to carry an array of multiple, flexible imaging equipment. Military needs included the classic strategic and tactical needs for cost-effective intelligence gathering and with round-the-clock surveillance capability.

Most promising was the considerable interest found among non-military governmental agencies and potential commercial users, both domestic and foreign. Applications such as border patrol to control illegal immigration, smuggling, drug running and terrorist activities, as well as search-and-rescue and disaster control, were cited as areas where unmanned vehicles could provide a cost-effective way of providing full-time large area coverage.

Commercial and scientific interests cited atmospheric monitoring, forest and crop surveys, forest fire control, fishing territory monitoring and protection, and similar activities which require long duration, inexpensive surveillance.

OK — THE STUDIES INDICATED not only a few, but many requirements for such a vehicle — and the next move was to turn the engineers loose to come up with a moderately low-cost bird to fill these needs.

Doug Fronius, experienced aeronautical engineer and pilot, was given the assignment of project engineer.

Initial configuration and three-view drawings were developed by Ladislao Pazmany, highly respected for his design of general aviation aircraft. The lead designer of the 410 then became Dave Harper, a veteran Teledyne Ryan engineer.

Design criteria were moderate initial cost, low operational cost, simple maintenance, all-weather day/night operation, conventional landing/takeoff, long range and endurance, high service ceiling and large load carrying capability.

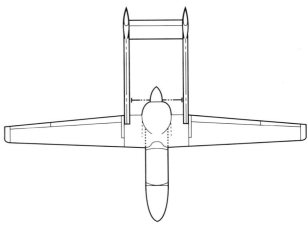

The prototype vehicle, which emerged from the plant in August 1987, was not only pleasing to the eye but built with no-nonsense commercial accessories wherever possible. It is comparable in size to many small single-engine civilian aircraft currently flying, and has a maintenance philosophy that any A&E mechanic can understand.

The 410 is a twin-tailed aircraft with a wing span of 31 feet and length of 21½ feet.

The pusher propeller and retractable nose gear enable unobstructed viewing by retractable sensors. Top speed is 190 knots with reduced speeds used to obtain maximum range or endurance. Service ceiling is over 30,000 feet.

Gross weight varies from 1,600 to 2,800 pounds depending on customer payload and endurance needs. The payload compartment has a volume of 24 cubic feet.

On the ground, one man can remove and stow the wings in a matter of minutes, reducing the vehicle to an 8-foot width that permits it to be towed on highways on a conventional trailer. Turn-around time between flights is limited to the time it takes to refuel and replenish any other stores required for the mission.

Unlike most UAVs, takeoff and landing can be accomplished on a 750 ft. unprepared strip. This avoids the costs and equipment used with air-launched and ground-launched systems, expendable boosters and recovery.

A major departure from the standard aluminum construction of most civilian aircraft is in the use of a high density structural foam layer between fiberglass epoxy skins, a technique used in building high-performance sailplanes. Wing spars and landing gear are of graphite-epoxy composite.

Power source is an off-the-shelf general aviation Lycoming 160 hp. turbo-charged four-cylinder gasoline piston engine, swinging a three-bladed prop.

Performance criteria for a long range, long-endurance surveillance aircraft underwent several analyses to determine best compromise for anticipated missions.

DESIGNED TO FILL THE GAP between mini-RPVs and long endurance *high-flying 'spy' planes, the Model 410 was rolled out at TRA's San Diego plant in late 1987.*

Dave Gossett

THE NO-NONSENSE DESIGN AND CONSTRUCTION similar to that used in high performance sailplanes, *including commercial accessories, were planned to make the new RPV truly affordable.*

Above: **THE UNUSUALLY LARGE PAYLOAD** bay provides 24 cubic feet *of space for multiple large-scale mission sensors.*
Below: **FOR QUICK TRANSPORT OR STORAGE, wings can be removed** *or stowed, reducing the vehicle to an 8-foot width.*

PRELIMINARY DESIGNS called for an 80-hour endurance with a 100 lb. payload, with growth capability of 300 pounds, altitudes of 45,000 feet with wingtip extensions and a higher compression turbo-supercharger. Eighty hours, with a minimum cruise speed of around 80 mph would mean staying aloft for more than three days.

As the design progressed and more realistic goals became apparent, revised mission capabilities call for sorties of at least twelve hours duration flown to maintain successive flights of the UAVs on station continuously. Routine launch and recovery operations are being considered for border patrol missions, since the 410 has sufficient endurance capability to permit takeoff in the early evening and remain aloft until after sunrise the following morning.

Despite its innocent exterior appearance, internally the 410 is capable of housing exotic equipment that has been the trademark of Firebee systems for decades past.

Two primary guidance modes are provided. In the command guidance mode, the vehicle responds to uplink commands issued either by the local or remote control operator. The autonomous guidance mode employs an onboard navigation system consisting of VLF/Omega or optional Global Position System receiver and navigational computer. The bird can be preprogrammed through a sequence of waypoints without operator intervention to perform preplanned over-the-horizon missions beyond the range of the ground control center.

It is designed to use ILS (Instrument Landing System) for automatic landing. Sensors will be housed in the large payload bay with the stabilizing systems. FLIR (Forward Looking Infrared) or daylight TV can be mounted in a retractable turret. The nosewheel retracts to give a clear field of view for the sensors.

WITH NO MILITARY specifications to delay manufacture, the first aircraft was ready for initial ground test in December, 1987. The airspace around the city of San Diego, with its complex Navy and airline traffic, was no place to conduct unmanned vehicle test flights. Across the mountains to the east of San Diego is the Imperial Valley desert town of Holtville and local airport operators welcomed the test crew.

Gene Motter, veteran of off-site operations for many years, headed up the seven man TRA team. Local operators, 144 miles east of San Diego, with tongue in cheek, named the airport Shangri La, but to Gene and the crew it provided remote isolation from San Diego's bustling air activity. Using ingenuity and improvision as key elements of their operation, they moved in with mobile support vans to begin taxi tests.

Landing an unmanned airplane on a conventional strip was no new experience for TRA, having flown many such flights with the Compass Cope-R birds. However, the flight control system on the 410 was untried and shortly resulted in a flight which damaged the landing gear. Back to San Diego and the drawing boards.

Redesign of the flight control system was completed and the crew was ready to return to Holtville when Bob Mitchell, TRA's new President, a former RAF squadron leader, called for a change of plans. An experienced pilot, Mitchell suggested that the balance of the R&D flights be conducted in a manned version of the bird. Consequently the automated unmanned system was temporarily removed and manual controls installed.

By May of 1988 the first manned flight was conducted. Ray Cote, veteran air race champion and TRA company pilot, put the 410 through its initial flights and by June the airplane was ready for additional tests.

The presence of a pilot aboard was primarily to explore the aircraft flight envelope and mechanical systems operation without the temporary complication of a flight control system. Succeeding Cote as company pilot is a gal named Denise Bright. Experienced, jet-qualified and possessing all of the ratings required for a top-notch executive aircraft pilot, dainty 'dollar-bright' Denise took over the remainder of the test program, flying the plane through nearly 40 takeoffs and landings.

"With me at the controls," Denise says, "the 410 was indeed an un<u>man</u>ned aircraft."

Satisfied that they had a configuration that would meet most of the definitized missions, the crew returned the 410 to San Diego while the marketing people worked to find an interested customer.

IN TYPICAL ROLLER COASTER fashion, the feast-or-famine syndrome of the aeronautical industry hit TRA in April 1988. The **Model 324** Egyptian program (see next chapter) was well under way and when the Navy/Marines/Air Force mid-range UAV requirement became known, the decision was made to enter that competition with the **Model 350**.

Converting the proven capabilities of the 324 into a new vehicle meeting the requirements for the Mid-Range UAV (MR-UAV) in

FOR INITIAL UNMANNED FLIGHTS the 410 was trucked to California's Imperial Valley *and reassembled for taxi tests.*

JET-QUALIFIED TRA executive pilot, sparkling Denise Bright, took over manned test flights *to better validate performance and equipment.*

the limited time allotted demanded the full capabilities of the engineering staff.

The Model 410 had just completed its first manned flight and was ready for further engineering development and testing when the demands for the 350 took priority over any additional effort for continued work on the long-endurance 'affordable' drone.

After a hiatus of over a year during which the 410 crew was employed on the '410B' project (see pages 177-181), and on the subsequent Model 350 mid-range vehicle, work resumed on the 'low cost' 410 under direction of Bob Hamrick as program manager.

For some time, TRA researchers had been analyzing how the cost of avionics and control system packages might be reduced since these items represent about 30% of the total cost in RPVs used for military missions.

Inasmuch as the flight test program in the manned version of the 410 had validated the aircraft's aerodynamic characteristics a double benefit could be achieved by installing the recently developed more advanced avionics system in the 410 airframe.

The 410 would be back in the air as an unmanned vehicle and the new avionics equipment could simultaneously be tested.

During the 1988 initial tests of the 410 at Holtville, the avionics then installed had not performed satisfactorily. To hold down UAV costs, commercial equipment had been installed but was not adequate for the stern requirements of unmanned flight.

The heart of the new avionics/control package is TRA's production Microprocessor Flight Control System (MFCS) computer. Its accompanying autonomous flight control system uses a commercially available GPS/Loran navigation system. Manual, remote control operation is also provided.

Now installed in the 410 for the dual-purpose test program is a more sophisticated mix of standard military (MIL spec) equipment and high quality commercial units. Reliability will be greatly improved and cost reduced.

Meanwhile, interest in the capability of an updated, cost-efficient 410 RPV continues to grow as new requirements both military and commercial are identified.

Victor Sotelo

LATE BULLETIN:

The rejuvenated Model 410 configured for unmanned flight, and with new avionics and control equipment, took off from the 'runway' at Cuddeback dry lake on the Edwards Air Force range for a successful first flight and landing on July 25, 1992.

EGYPT'S 324 SCARAB

SHERMAN LINEN for Teledyne Ryan Aeronautical

IT IS PARADOXICAL how a vacation trip abroad — to get away from some of the burdens that go with being a key business executive — often leads instead to new business opportunities.

Ask G. Williams Rutherford, retired Teledyne executive. Though he had frequently visited Israel as a Ryan Aeronautical executive and President of that company's Continental Motors subsidiary, he had never been to Egypt, land of pyramids and the sphinx.

In mid-April 1983, Bill Rutherford was in Geneva accompanied by his wife, Anna Gwyn, and their daughter Amanda. Why not go on to Cairo before stopping over in Israel?

A certain amount of diplomacy and tact had to be considered.

KARSH, Ottawa

G. WILLIAMS RUTHERFORD

"**D**URING THE TIMES I had gone to Israel in the late 1960s," Rutherford explained, "when they were buying Continental engines for their tanks, officials there asked me not to visit any of the Arab countries. They never said that it was a condition of doing business with them but in those days the situation between Israel and the Arab countries was a lot more tense than it was in the '80s.

"At that time I made the decision that I would go to Israel on business trips and have other Continental executives go to the Arab countries. Consequently in 1983 before going on vacation I talked with my friends in Israel reminding them that they now had a peace pact, an open border and airline service between the countries, and would they mind if my family and I went as tourists to Cairo.

"They said, 'No, go ahead.'

"From Geneva, we flew to Cairo and were met on arrival by the principals of Sahara Overseas Services, representatives in Egypt of Teledyne, Inc. Retired Egyptian Generals El Sayed Aly Nadim and General Sayed

Maged Aly, accompanied by their wives, were our hosts for the next two days.

"They arranged our accommodations at the Mena House in one of the most beautiful suites I have ever seen. I had told them we wanted to just be tourists; that I didn't want to do any business.

"That evening we had a long dinner in the diplomatic quarter in Cairo, then all returned to the Mena House where the Generals and their wives also stayed overnight so we could be together in the morning, as our hosts had planned a very heavy schedule for us.

"One of the first sights we saw was spectacular. When I woke up in our hotel room I pulled the drapes and I was standing there looking right at the pyramids. I'll never forget that moment as long as I live. That night we all went to the sound and light show at the sphinx.

"The next day it was the bazaar, a short trip and lunch on the Nile, and then in late afternoon we changed hotels – to Heliopolis near the airport – as we were leaving at midnight for Tel Aviv.

"After checking in at the Sheraton, the Generals asked would I please join them in a visit to their office. It would have been inconsiderate not to have agreed since they looked a little downcast when I repeated my earlier statement that we weren't in Cairo to do business.

"Almost reluctantly I agreed. General Nadim mentioned that his daughter ran the Sahara Overseas office for them, so our daughter Amanda accompanied us to their office, which was nearby.

"The girls struck up their own friendship in the outer office and we went into one of the other offices.

"General Nadim said, 'Bill, isn't there some kind of new business we could originate, through Ryan, so we could be more useful in creating a viable project for Teledyne?'

"That caught me somewhat off guard, so on the spur of the moment I said, 'Sir, why don't you offer your government a drone system?' They said, 'We've heard something about them. Why don't you tell us more.'

"**H**AVING BEEN INVOLVED with Ryan drone systems since joining the company in 1951, it was easy to respond to their request for more information. I told them about drones as aerial targets which could test the effectiveness of their defense systems. How they had been operated on reconnaissance missions during the war in Vietnam. How they could meet the needs of many types of unmanned missions without putting pilots of manned aircraft at risk.

"ONE OF THE FIRST SIGHTS WE SAW WAS SPECTACULAR. When I woke up and pulled the drapes *I was looking right at the pyramids. I'll never forget that moment as long as I live."* *Rutherford's picture is of the Great Pyramid of Cheops at Giza.*

"They were fascinated and asked many probing questions.

"'A drone is like a pickup truck,' I explained. 'If I sold you a truck I wouldn't want to tell you that you could carry crates of fruit and vegetables or sacks of cement, or anything like that. You would make up your own mind what you were going to use it for.'

"'I really believe that is what a drone system is all about. **It gives a country that has a complete drone system a lot of options.** One of the best is reconnaissance of your neighbors so your intelligence people really know what is going on.'

"I of course was aware that the Egyptians knew about the Ryan 124I Israeli reconnaissance drone operations in the 1973 Yom Kippur War, so I wasn't dealing with people who were hearing about drones for the first time.

"We discussed things like could they build drones there after we had designed it and built an initial quantity. I assured them we could design a system for them that was basically of Kevlar and other non-metallic materials, and could provide molds and assembly tooling so they could eventually put the drone into production without much effort.

"We talked about navigation systems and launching methods, including a truck for ground-launched missions. Also, we discussed air launching, and about recovery methods, and all the things that make up a complete drone system.

"They were very interested; made some notes, and asked me how much the start-up cost might be. I said the initial program might be around $40 million to design, test fly and produce a small number of vehicles.

"After this intense conversation of about an hour we all went back to the Heliopolis Sheraton Hotel, had dinner, thanked our Egyptian hosts and they put us on our midnight plane to Tel Aviv."

"ABOUT THREE WEEKS AFTER our return from the Middle East," Bill Rutherford recalls, "I received a phone call at home from General Nadim on a Sunday afternoon. He was very excited. 'Bill,' he said, 'a close friend of ours is an aide to Field Marshal Mohammed Abdel Halim Abu Ghazala. We gave him a long briefing about the drone system we discussed and he has gone over it with the Field Marshal. I think we have a sale!'

"To say the least, I was somewhat stunned.

"'That's wonderful,' I told Nadim. 'So what do we do now?' He said, 'I think we should get started on drawings, and get some people over here.'

"Well, that's exactly what I had in mind; so we decided to get going and get the job under contract.

"At that time Marshal Abu Ghazala was not only Minister of Defense, he was second in command in the Egyptian government. If he wanted the project it was going to go, because he set the priorities on military projects.

"We put a small team of experts together under Gene Timmons, got the design started on paper and began exchanging visits with the Egyptians.

"By the time of the Paris Air Show two months later, in June 1983, Timmons made the final briefing to Marshal Abu Ghazala at the Teledyne chalet."

THE MONDAY AFTER General Nadim's surprise phone call from Cairo, Rutherford called a number of technical and marketing people into his office.

"Guess what!" Rutherford began, "I think we can sell some drones to Egypt, if they can do what I outlined to them during my recent trip. It's a hurry up job. Why don't you guys get to work and come up with the design? I want you to be able to brief Marshal Abu Ghazala, who will probably be at the Paris Air Show next month."

Responsibility for the new project had been given to Gene Timmons, at that time in charge of advanced systems, including drones and international programs. "The advance design group was headed by Oliver Cathey and aerodynamicist Arnie Otchin," Timmons explained, "and we benefitted by strong support from Bill Grenard, Vice President-Engineering, who was a very positive player."

"We knew that we had to have some new technology to offer the customer, but not so new that we would be treading on classified areas. That would cause understandable concern in the State Department and Munitions Control, and could hold up an export license.

"So, we had to come up with design features based on available technology but that is new for third world countries. That was a paradox we had to deal with."

Because the Egyptian Air Force had no planes capable of air launching UAVs, or existing ground-launch installations, it was obvious that TRA would have to supply a very complete, fully integrated system which would include not just vehicle and propulsion but also transportable launch, control and retrieval units.

The basic airframe manufacturing solution was to consider use of composite materials instead of conventional metal structures for the new RPV.

Over the years, TRA had accumulated considerable experience in building components of composite materials. Such structures use reinforcing materials and moldable products including fibers such as DuPont's Kevlar, glass, boron or graphite.

Composites are usually stronger than an equivalent weight of metal and are thus very attractive as a structural element, and are more resistant to corrosion. Composites are easily formed to complex curvatures and shapes, and one large structure often replaces several conventional metal parts.

In manned aircraft, for example, TRA had long provided graphite and Kevlar composite components such as those used in TRA-built airframe structures produced for the Army's AH-64 Apache attack helicopter, as well as for Firebee components.

Thus, in designing the new RPV for Egypt, Kevlar reinforced epoxy was selected for the fuselage and major flight structures.

PRELIMINARY LAYOUT (of the then Model 124RE) showed a wingspan of 12 feet and length of 20 feet. Range was to be 1,400 nautical miles, payload 250 pounds, speed Mach 0.8 and service ceiling 43,000 feet.

A 970-pound thrust Teledyne CAE 373-8C engine, derived from that firm's J-402 Harpoon power plant, was selected to power what the Egyptians were to call the **Scarab**, and was to be mounted atop the fuselage. Take-off from the launch trailer would be assisted by a jettisonable booster rocket.

To the naked eye, there appears to be little similarity between the narrow-span **Model 324 Scarab** and the earlier, long-wing Model 154 Compass Arrow and Model 235 Compass Cope RPVs, but listen now to Doug Fronius, later engineering manager:

"The most important design features of the Scarab including its external configuration, were established through experience gained in the successful high-altitude, long-endurance Compass designs.

"The 324 had to be scaled down to a mid-range vehicle to meet mission requirements in the Middle East.

"Retained design features included fuselage shape and cross section with slab-sided, flat bottom surfaces. Engine placement, including inlet and exhaust, again atop the fuselage. Wing and tail section location including twin vertical fins also followed the Compass Arrow style."

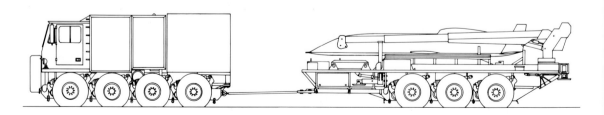

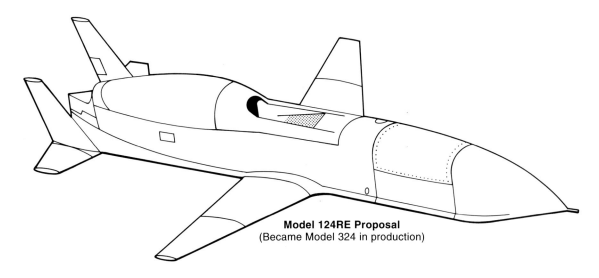

Model 124RE Proposal
(Became Model 324 in production)

Even with a fully loaded gross weight of 2,400 pounds, the new aircraft was a diminutive small brother to the full scale 154 surveillance vehicle.

In-flight recovery following completion of the mission is accomplished by a two-stage parachute system. Among unique features are the air bags which deploy from the bottom of the vehicle to absorb impact forces on touchdown.

ARMED WITH BRIEFING CHARTS and engineering data, Timmons and his team arrived in Paris for the opening of the Paris Air Show, May-June of 1983. There they met Generals Nadim and Maged of Sahara Overseas.

"They made arrangements," Timmons said, "to get Field Marshal Ghazala into our chalet. He came with his staff of ten or 12 people and we briefed him for about an hour on the concept and preliminary details for an Egyptian tactical reconnaissance RPV system.

"Despite the number of staff officers present it was a very personal briefing for the Field Marshal. At the conclusion of the briefing he turned to his staff and said, 'That's what I want. This is what we need to do!' Turning to me he said, 'My staff will make further contact with your company.'

"That started the whole ball rolling."

A steady exchange of TRA specialists and Egyptian officers between Cairo and San Diego brought a consensus which eventually culminated in a formal commercial contract between the Arab Republic of Egypt and Teledyne Ryan. Ollie Cathey teamed up with W.R. (Bill) Hirst, contract business manager, to keep things moving in Cairo. Then the final negotiating session was presided over by TRA president Steve McCarter.

"THIS WE NEED TO DO"

During negotiations the Egyptians operated with two teams. Timmons told how they "go through all the technical aspects first, not discussing contract terms at all. When they are satisfied with design matters they bring in a whole new team and start the contractual negotiations. We went through both of these phases with them."

The technical conferences went on for over six months. One Egyptian group led by Col. Dr. Hafez M. Zaki was based in San Diego for several years.

"By this time," Timmons says, "we had transitioned prime responsibility to Norman Sakamoto, a veteran program manager, and moved the project into a secure 'skunk works' controlled access area.

In summer 1985 the Navy issued an RFI (request for information) for a mid-range vehicle referred to as midi-RPV, and Teledyne Ryan early on offered the basic 324 design as a candidate. This Navy interest preceded by several years formation of the new Joint Project Office, formed in 1988, to coordinate all military RPV programs. The 'midi' project then became the new MR-UAV program in which TRA later participated.

"**T**HE SCARAB PROJECT is unique," says Fronius. "It is the first *complete* RPV system we have supplied any customer. System is the key word.

"We were the prime contractor/designer including the aircraft, payload integration, ground control station, truck to move it around, the trailer on which to launch it, the launcher, booster system, command and control system.

"All of these components were brought together in a complete self-contained turn-key system, plus all the ground control equipment necessary to do it, right down to the hand tools, cans of oil, rags ... everything you need to maintain and operate this unmanned vehicle we were to deliver to the Egyptians.

"You could drop all the boxes we delivered in the middle of the desert and with no other equipment, not even power because we also supplied the generators, you could operate an RPV. This is the first system we have done so completely."

AFTER CONTRACT SIGNING in June 1984, work began in the shop, also under Cathey's leadership, to build the first full-scale models of plaster and fiberglass in order to study production methods and determine air flow and potential flight characteristics.

WHILE INITIAL SHOP WORK was under way, *aerodynamicists were busy at the Vought wind tunnel checking air flow and flight characteristics of a scale model of the new bird.*

Engineering's Bill Grenard was already working with Burt Rutan of Scaled Composites, Inc. of Mojave, California, on composite structures being considered for TRA's long-endurance 'Spirit' project.

"Soon after we signed the actual contract document," Timmons recalls, "a decision — an economic one — was made to expedite production by having Rutan build the first composite units."

STRUCTURAL INTEGRITY OF THE FULL-SCALE composite airframe *was fully validated in the TRA vibration test laboratory.*

Ryan already based some operational people at Mojave, because of its proximity to nearby Edwards Air Force. Rutan was well experienced in building composite structures for aircraft and was also working on his projected round-the-world Voyager nonstop flight on a single tank of fuel. For that unique project Teledyne Continental was supplying specially-designed water-cooled piston engines.

When it came time for the actual production run of 29 Scarab vehicles, Scaled Composites became the major airframe subcontractor.

Other major suppliers, under TRA design considerations, provided command and control systems, inertial navigation, booster engines and the unique mobile launch and recovery tractor/trailer vehicle.

The intelligence-gathering payloads, including high-resolution cameras, were selected and contracted for separately by the Egyptian government, but TRA technicians integrated them into the overall reconnaissance system.

EAST IS EAST; West is West — and for the twain to meet took a considerable amount of patience, understanding, forbearance and a willingness to see the picture from the other side of the fence. Negotiating a military contract with the U.S. Department of Defense is quite different from closing a deal with an Arabic country.

Psychologically, business ethics and procedures which might raise eyebrows in Western countries are accepted as a normal way of operating in many of the Middle East countries.

Westerners often mistake the Caucasian appearance of Mideasterners as an indication of Western ways of thinking. However, in most instances, their thought processes are different. Often, the impatient proddings of a Western representative can create delays, disharmony and frustration on both sides of the negotiating table.

Ethics and procedures aside, the language barrier can lead to misunderstandings, and even a different calendar can cause confusion. With 'Sunday' falling on Friday in most of the Mideastern countries, a Yankee finds it difficult to say 'TGI Thursday,' instead of TGIF, and with Saturday as the start of the work week and first of the month starting on the 21st, business hours and days take more than a little getting used to. Add a ten hour jet-lag and time zone difference, communications are unreliable at best.

Norman Sakamoto, TRA's choice to take overall responsibility for the Egyptian contract, had excellent assistance on his negotiating team with Bill Hirst heading up that effort. With Bill to handle the contractual side of the business and Oliver Cathey responsible for the technical aspects, their sagacious approach to successfully negotiating a contract overcame most of the difficulties.

Three round trips between San Diego and Cairo covered a time span of over three months to reach a preliminary understanding. Despite the potential problems for both parties, progress was marked with the 'two steps forward, one step back' philosophy, and eventually a mutual understanding was reached. Assisting Bill and Oliver during these periods was Bob Scurlock, TRA vice president and top contract administrator, Dave Hall and eventually Hudson Drake who had replaced Steve McCarter as TRA president.

Principal contact and go-between was Sahara Overseas, the consulting firm headed by Egyptian ex-generals Nadim and Maged, who also represented several other Teledyne companies in Egypt. The value of a reliable in-country representative firm cannot be overlooked. With that firm's understanding of Egyptian business practices and their ability to work with the TRA negotiating team, many of the rough spots in the parleys were smoothed over.

WITH CONTRACT TERMS finally agreed upon, an informal debut of the project was disclosed to a special audience in Cairo late in 1984. Teledyne was displaying a one-quarter scale model of the 124RE, with Egyptian Air Force markings, at the Defense Equipment Exhibition at Alamaze Air Base.

Among those inspecting the model of the new mid-range Scarab tactical reconnaissance system was Egyptian President Hosni Mubarak, accompanied by Field Marshal Abu Ghazala.

Dave Gossett

IN ONE-QUARTER SCALE MODEL OF NEW RPV design, shown in Egyptian markings, *TRA engineers sought the best combination of desirable characteristics.*

G. Williams Rutherford Collection

INTERESTED SPECTATORS AT DEFENSE EXHIBIT in Cairo *were Egyptian President Hosni Mubarak, right, and Field Marshall Abu Ghazala.*

Three years later the first full-scale Model 324 Egyptian RPV was displayed at the international military equipment exhibition; this time with TRA President Hudson B. Drake and Vice-President Norman Sakamoto present, accompanied by General Nadim of Sahara Overseas.

IN ANY BUSINESS ARRANGEMENT, especially those involving foreign governments and military products, there are always a lot of factors at work. Working your way through the maze of regulations, foreign and domestic, and political considerations, is a very tricky course.

As G.W. Rutherford, who was Teledyne's Group Executive overseeing TRA and other defense-oriented companies, explained, "We were faced with two concurrent problems. One dealt with financing the contract. The other was difficulty with the export license from Munitions Control, the State Department and the Pentagon.

"Hudson Drake had successfully launched the Egyptian job inside TRA; he'd gotten it up and going. I never went back to Cairo, but Hudson and Chuck McGill, TRA vice-president, later made many trips there. They directed company strategy to keep the program coordinated between the company and the Egyptian government.

"When there was an interruption in summer 1988 in the planned schedule, Hudson, by then Senior Group Executive of Teledyne, Inc., personally undertook the task of straightening out both the export license and the payment problem. He rates an A+ for performance in getting the program back on track."

HUDSON B. DRAKE
Senior Vice President, Teledyne, Inc.

His Teledyne associate gave
him an A+ for performance.

THE EFFECTIVE DATE for the TRA/Egyptian contract was 29 October 1984, calling for the delivery of 29 RPVs, three launch and recovery vehicles, three sets of ground support equipment, and tooling. Seven Egyptian representatives took up residence in San Diego to monitor the program.

Specifics for the bird's capabilities included medium altitude unmanned reconnaissance, low radar signature and observables, rapid flexible response, payload modularity/growth, exportable technology with co-production, and the development of a mutual Quick Reaction Capability.

With such a broad base from which to work, Norm Sakamoto's job was a challenge.

Co-production was not new terminology at TRA. Earlier, arrangements had been made with the Japanese to share in the manufacture of the Firebee target, and Norm appointed Brian Eagan and Dave Hall to investigate the Egyptian manufacturing capabilities.

After a two-week sojourn in Cairo, Brian reported, "The Egyptian facilities were truly amazing. The facility intended for possible 324 production was a building built originally for the manufacture and assembly of a British helicopter.

"At a cost of over 250 million British pounds, it contained its own power source, complete air conditioning, electronic areas and a paint booth capable of handling an aircraft of almost any size. I don't know why the British deal fell through the cracks, but it was apparent that if the Egyptians so desired, they certainly had the personnel and the equipment necessary to handle an aircraft the size of the 324 with ease."

Prior to the award of the contract, Gene Timmons had indicated that the design of the new bird would have to incorporate some new technology but not so far out as to cause conflict with any classified areas. As a consequence, the 324's exterior appearance was drastically different from the standard Firebee, but resembled a preshrunk version of the Compass Arrow vehicle. With the engine mounted above the fuselage, IR detection was lessened, and with canted sides to the fuselage, radar acquisition was diminished.

ALL-TERRAIN LAUNCH AND RECOVERY (LRV) truck and trailer *combination, operated by three-man crew, provides total mission support.*

"EVEN THOUGH 14 MAJOR subcontractors manufactured items for the 324," explained Fronius, "TRA had complete responsibility for design. For example, we did all the design work for what was added to the basic chassis of the mobile Launch and Recovery Vehicle (LRV).

"The LRV builder for us was Standard Manufacturing Company of Dallas. We selected the command and control shelter and positioned it on the eight-wheel tractor, as well as designed the superstructure of the six-wheel transporter. TRA also designed and built the launch rails and added them to the transport/trailer chassis."

The all-terrain truck and trailer combination provides all necessary mission support functions for launch, in-flight control and recovery, and can be operated by a three-man crew

Turned by differential rotation of the wheels, the tractor has a top speed of 52 mph and range of six hours. Thus, the RPV system can be operated from fixed-base or remote, unprepared sites. Mobile field operations permit 30-minute set-up and five minutes preflight time.

Paul Bunner, the ingenious creator of the first mobile launch system for the Firebees in the early '60s, eventually ended up manning the LRV, both during the flight test cycle and later with on-job-training (OJT) for the Egyptians. The elaborate new system was a far cry from the surplus flat-bed trailer that Paul scrounged at White Sands 20 years earlier to prove the mobile zero-length launch system for the Army.

Left: CAMOUFLAGED TRACTOR/TRAILER LRV carrying its Model 324 RPV *makes a trial run over the Mojave Desert..*
Below: AN EGYPTIAN AIR FORCE C-130 transport visits Mojave *to test loading of LRV unit.*

THE FLIGHT TEST PROGRAM was to be conducted at the Mojave operations base; and the selection of Burt Rutan's Scaled Composites company at Mojave proved to be fortunate in several respects.

Originally that company was to build only two ship-sets of components, but their ability to respond to changes with a minimum amount of confusion and wasted time resulted in transferring major manufacturing to the desert facility. A team of fifteen Ryanites under the direction of Dick Thomas eventually finished all 29 of the birds in the Mojave Airport Hangar facility.

Sakamoto gathered the cream of TRA's field service personnel to manage the flight test of the new UAV. Manning the outfit started in August '85, and eventually encompassed veterans Paul Bunner, Larry Hurley, Gene Juberg and Engineering Manager Billy Dickens. Selection of the most outstanding talent in the drone operation business proved to be the key to the success of the flight test program.

Assembling and testing all of the components going into an untried airframe was a monumental chore. Engine selection for the bird was the Teledyne CAE modification of its Harpoon turbo-jet with a 970 pound thrust, augmented by the Thiokol rocket motor booster.

TRA GROUND CREW CHECKS ALIGNMENT of 324 vehicle *for initial launch at operational base adjoining Edwards AFB.*

Dave Gossett

SCALED COMPOSITES, INC., WORKING WITH TRA pro-duction and engineering veterans, *produced the initial order of Scarab vehicles for Egypt.*

Ironically, it was the Harpoon project that influenced cancellation of an earlier Ryan program for the strike version of the Firebee, but with the sometimes invent-build-cancel cycle of military procurement, the modified engines found a home in the Model 324.

The electronic guts of the new bird would have amazed and confused the designers of the standard Q-2C Firebee. The 324 used a C-band system furnished by Vega Precision Laboratories for flight and command control. Litton Industries INS was used for stabilization and guidance, updated with a Rockwell Collins C/A code Global Position System (GPS) for navigation with pinpoint accuracy assured.

The flight control system incorporated the TRA designed Mission Logic Control Unit (MLCU) which uses microprocessor-based logic to control speed, payload, guidance and navigation, propulsion, fuel, electrical and recovery systems. With all of this gear packaged into a volume smaller that a fifty gallon drum, the new bird was truly an advanced state-of-the UAV art.

BEGINNING IN MARCH 1987, the year-long contractor's flight test program at Mojave accumulated 18 flights with more than 16 hours of airborne operations. Except for two night flights, all were conducted on Saturday and Sunday, a restriction that tightly limited operations at the desert operations base. As usual, and despite supervision by the experts, all flights were not textbook perfect.

The second flight, conducted in April, resulted in a bird being expended when the booster exploded during the launch phase. A two-second duration flight was somewhat disheartening, but it was one of the only two birds lost in the entire Mojave test program.

166

UP, AND AWAY AT MOJAVE! As 324 is launched *from its transport trailer, flight control is assumed by the command station housed on board the system's truck unit, right.*

A 44-minute flight on February 28, 1988 was the culmination of four years of dedicated effort by the entire 324 team. Norm reported, "From launch to recovery, it attained all objectives with precision." Completely hands-off, the airplane-without-a-man-in-it, covered a thousand-mile track at Mach 0.8 and returned to a pinpoint recovery area with all essential electronic surveillance gear doing the job for which it was designed.

Delivery of the 29 birds, ground support equipment and spares began in October, 1988. By April the following year, a four-man team initiated the site activation in Egypt. Under the leadership of Larry Mesecher, 13 weeks of on-the-job training was given to the Egyptian technicians. Site selection finally centered on an area roughly 70 kilometers south and west of Cairo in a district phonetically called Kom Oshiem.

ON COMPLETION OF MISSION, the onboard two-stage parachute *system is activated and air bags beneath fuselage are inflated to absorb landing impact. Vehicle is then replaced on trailer for return to base.*

167

Jack Broward

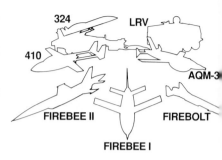

324   LRV
410   AQM-3
FIREBEE II   FIREBOLT
FIREBEE I

NIGHT LAUNCH OF 324 SCARAB, one of two conducted *during flight test program, leaves a trail of fire across the desert skies.*

FOR 1988 AIR/SPACE AMERICA EVENT, TRA displays a full-hand of its RPVs. *From 324 Scarab vehicle and tractor-trailer unit, at top, other vehicles are, clockwise: 147 Lightning Bug, Firebolt, subsonic Firebee I, supersonic Firebee II and Model 410 affordable RPV.*

Dave Gossett

OPERATIONS IN EGYPT

TECHNICIANS ASSISTING EGYPTIAN personnel have apartments in Cairo, largest city on the African continent. TRA representatives travel to the Air Force base at Kom Oshiem by rental cars with Egyptian drivers. Their route takes them past the pyramids at Giza.

Operations are conducted from the Kom Oshiem base; others are staged from locations a hundred kilometers or more northwest into the desert near Wadi an-Natrun.

Equipment, supplementing the Launch and Recovery transporter and trailer, can be flown there by Chinook twin-rotor helicopters.

Later, in case they are required, the Chinooks are available to recover the 324s after their descent by parachute at the end of the mission, and transported back to base.

At wadi (dry watercourse) desert locations, a back-up unit with an RPV aboard its trailer is available and ready to be launched in case the primary bird encounters any technical delay.

Operationally, the Egyptians use the original 124RE model designation for their Scarab RPVs.

Ray Zelinski, TRA's in-house support manager for the on-site program, logged all of the flights in Egypt with the first one occurring on September 9, 1989. Remarks: EXCELLENT FLIGHT.

However the entry for the second flight reads: Excellent Flight — Chute Failure — Loss of Vehicle. This was the only bird expended during the course of a ten-month period in which 15 proving flights in Egypt were made. Max endurance occurred on January 16, 1990 with an hour and fifty-five minutes logged and the remarks: Excellent Flight and Recovery — No Damage to Vehicle.

Minor damage occurred on several flights due to high winds on recovery, but the fact that a three-man crew could accomplish launch, flight missions and pinpoint recovery spoke well for the training that went into the program. Under Paul Bunner's supervision, 12 Egyptian control operators were trained in TRA's simulator lab in San Diego in May.

Additional successful operations were conducted in 1990 from Kom Oshiem by Egyptian technical personnel. With the 25th flight, the Scarab system was declared operational with the Air Force. To that point, there had been only the one flight in country which resulted in loss of an RPV.

Much of the credit for getting Egyptian flights up to full operational capability was due to Larry Mesecher, Lloyd Morrison, Sherman Linen and other members of the TRA technical team who remained on site after the initial period of proving flights and training of Egytain operations.

As team leader and site activation manager, Mesecher was held in the highest regard by both Egyptian Air Force officers and their RPV personnel with whom the TRA people worked.

'Scarab', the name the Egyptians gave the 124RE (model 324), describes the traditional stone talisman carved to represent a beetle regarded by the ancients as a symbol of resurrection and immortality.

[By coincidence, the 15th century French word for this beetle is Scarabee - not too remote from Firebee - Ed.]

As a huge step from desert warfare with camels, the utilization of a twentieth century Scarab UAV is truly a resurrection for the Egyptian military.

And incidentally, the Scarab proved to be a turning point for TRA's drone business. In March of 1990, Bill Hirst, Chuck McGill and Bob Smith, executive director, were successful in negotiating a follow-on contract with the Egyptians for an additional quantity of Model 324 RPVs.

Development of the Model 324 and its successful introduction to operational status by the Egyptian Air Force served to jump start Teledyne Ryan's successful bid to design and build the U. S. Mid-Range UAV.

Bill Rutherford's vacation trip to Egypt in 1983 has paid remarkable dividends to TRA's UAV story.

From the palm-lined gardens of Cairo and the river Nile, where feluccas with triangular sails ply their trade, the technician's daily journey ends on the arid desert at Kom Oshiem. En route to the Air Force base, the famed Mena House at Giza is passed, with the Great Pyramid as a backdrop. Natives and their burden-laden donkeys are a frequent sight just outside the air base.

Below and opposite page: Linen's sequence camera records first Scarab launch from Wadi en-Natrun. Second 324 'bird,' far left, is on stand-by.

Above: Close-up view of Teledyne CAE 'Harpoon' rocket motor which boosts Scarab during launch phase. Beneath nose section is window for infrared camera.

At remote base, a flight-ready 324 is off-loaded from Chinook helicopter and will be transported to LRV for launch.

Above: With attenuation bags deployed, the 324 is in tow to base operations beneath Chinook helicopter.

Below: On landing by parachute, this RPV dug a wing into the desert sand. But, no problem; no structural damage. Three days later the bird is ready to fly again.

Egyptian launch (LCO) and remote (RCO) control operators team.

Col. Mohamed Rashed Reda, Egyptian chief RPV operations engineer, center, with visiting TRA personnel. From left, Larry Mesecher, team leader; Chuck McGill, vice president; R.A.K. Mitchell, president; and Bob Smith, executive director.

KEY OFFICERS OF THE EGYPTIAN AIR FORCE RPV Regiment and TRA technical experts. *From the left, Col. Reda (chief engineer), Sherman Linen, Larry Mesecher (TRA team leader), Col. Sharif (commander), Lloyd Morrison, Col. Osama (vice commander) and Maj. Fekry (payload chief).*

Doug Fronius

RANKING EGYPTIAN OFFICERS concerned with the Scarab RPV program. *From the left, Col. Ashraf, Gen. Mokhtar, Col. Wael, Gen. Hafez M. Zaki (project manager from the beginning) and Gen. Abdallah.*

Left, and above: All-Egyptian launch and recovery crews preparing Scarab RPV for its next flight. A confident crew poses **below** with their bird now ready on the LRV platform.

Joe Janes

BQM-145A PROJECT

Ssgt. MARTIN E. HAMILTON-U.S. Air Force

JANUARY 1, 1988 a new President took over as head of TRA. To knowing old-timers it was reassuring to again have a pilot-engineer as chief executive of the company established 66 years earlier by another airplane-oriented leader — pioneer pilot and aircraft innovator, T. Claude Ryan.

Robert A.K. (Bob) Mitchell, by good fortune, also had the same kind of 'can do' quick-reaction program management style as those who led the Ryan team which, on short notice, designed, built and flew unmanned Ryan reconnaissance jets on intelligence-gathering missions in the Vietnam war.

A former Royal Air Force squadron leader and RAF College graduate **aero***nautical engineer, Bob Mitchell received a master's degree in* **astro***nautical engineering at the U.S. Air Force Institute of Technology. Later, he emigrated via Canada to the United States where he was Vice President of space programs at Teledyne Brown Engineering before joining TRA.*

Dave Gossett

ROBERT A. K. MITCHELL

At TRA, Mitchell was replacing Hudson Drake who had recently been promoted to Vice President of TRA's parent company, Teledyne, Inc., of Los Angeles.

On the MR-UAV (Mid-Range UAV) program, TRA started out having to play catch up. It would have to compete with two powerful groups which already had multi-million dollar study contracts for designs which would meet customer requirements. It would take an all-out effort if TRA was to have any chance — an outside chance at best — of winning.

WHAT WAS TO BECOME the Mid-Range UAV program began life under the cumbersome designation JSCAMPS — Joint Services Common Airframe Multiple Purpose System. JSCAMPS was to **meet the needs of the Navy, Marine Corps and Air Force for a common tactical air- or ground-launched vehicle which could fly both aerial reconnaissance and aerial target missions.**

Priority would be given to the reconnaissance quick-reaction capability for pre- and post-strike high quality imagery of heavily defended targets.

The practice of individual military services pursuing, procuring and developing unmanned aircraft without benefiting from interservice exchange of requirements was inefficient and expensive. Through the new Joint Project Office (JPO) the services combined to take advantage of the mutual pool of available knowledge and technology.

Planned for the government's fiscal 1990 year were four major programs: close-range, short-range, medium-range and endurance UAVs. The medium-range bird was to be the first procured. TRA's success in landing this contract in June of '89 is a repeat story of aggressive effort, can-do attitude and a tremendous storehouse of technology and experience in developing and operating unmanned vehicles.

Prior to the establishment of the JPO, Northrop and Martin Marietta/Beech teams had each received $3 million contracts in 1987 for design studies for a medium range tactical air-launched vehicle. And each had flight-proven hardware which closely approximated the requirement.

FREEZING ALL UAV FUNDS pending formation of the JPO halted these preliminary design studies, but TRA felt it was in no competitive position to bid on the MR-UAV requirement. With its Egyptian Model 324 demanding full engineering talent and the company-funded Model 410 in work, few discretionary funds were available. Further, the air-launch requirements for the proposed design could not be met with the 324 vehicle, which had been developed for ground-launch only.

Still, in the opinion of long-time Ryan business development Vice-President A.C. (Tony) Richards, the prior decision to 'no bid' the MR-UAV project ought to receive a final management review.

Thus, a month after Bob Mitchell's arrival at TRA, the new president called into conference vice-presidents Richards; Norman

Sakamoto, head of the 324 program; engineering chief Gene Dotson — and Doug Fronius, program engineer on the affordable 410 drone.

What TRA had going for it in early 1988 was the fact that the 324 had been flight tested, with deliveries to begin in June, and that a derivative of that vehicle appeared feasible. Also, a target variant might also be developed from the same basic RPV design.

The financial and technical investments already made on the 324 project would make possible a greatly accelerated transition of a new bird from development to production.

The design and construction of the flight-proven 324 could be scaled down to meet mission requirements including air launch from four specified tactical aircraft including the F-18 and F-16. Major subsystems carried over from the 324 would have the advantage

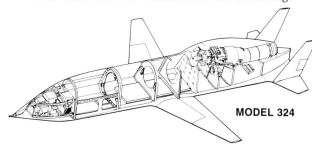

MODEL 324

of prior development, qualification and flight testing.

Development of the 324 had been divided between TRA's San Diego plant and facilities for assembly and flight test at Mojave, adjacent to Edwards Air Force Base. A similar routine could be used for development of the MR-UAV.

There were risks in playing catch up, but the odds were not impossible. However, time was of the essence. Response to the RFP — Request for Proposal — was due in two months.

The Model 324 mid-range reconnaissance system for Egypt successfully completed its test program of 18 flights on March 12. Two days later Bob Mitchell advised the Naval Air Systems Command (NASC) that TRA's candidate for the Joint Medium-Range Remotely Pilot Vehicle/Target Program would be a derivative of the 324.

"This approach," he wrote, "provides a benefit to the United States from the investment of over $60 million by TRA and Egypt, and will enable a highly-accelerated transition from development to production."

Work immediately got under way on a major proposal effort directed by Gene Dotson, Vice President-Engineering, with Frank

Oldfield and his people heading up the technical sections.

But would TRA's quick response under Mitchell's leadership win the contract?

Learning of the last-minute entry of the 324 derivative, TRA's competitors could not envision how the San Diego company had even a remote chance of being a serious challenge to their head start.

However, Mitchell saw a way to stage a deceptive flanking attack to outmaneuver the 'enemy.' If not that, at least a late fourth-quarter 'Hail Mary' pass that might score a touchdown.

THE '410B' PLOY

AT HIS STRATEGY meeting, Bob Mitchell told his associates, "The requirement is to launch from a tactical fighter and it's perceived that TRA does not know how to do this as the 324 is capable only of being ground launched.

"What we need to do to win is to launch an aircraft representative of our proposed **Model 350** off of a fighter — and do it on our own money as part of our proposal."

Doug Fronius picks up the story:

"Everyone except the 410 team went to work on the proposal effort, but 410 types were also available to help.

"Mitchell asked if I thought we could build a Proof-of-Concept vehicle; how long would it take, how much would it cost, could we fly it off an F-4 Phantom launch aircraft, and in the necessary time frame?

"That was on a Friday. I put together a briefing over the weekend, presented it Monday morning and at 2:30 pm that afternoon, June 22, 1988, received a go-ahead.

"The technical team which had just completed manned test flight of the 410 was called back from Holtville. They moved back into the 410 hangar at San Diego under tight security rules. By Wednesday they had started work on the **Model 410B**."

410? Yes, 410B.

AN EARLY TASK ON '410B' was fitting **CAE jet engine** *into the fuselage.*

Apparently the company on short notice had received a classified contract for some updated RPV based on 410 technology. Such was the word around the plant, where all the paperwork called out materials and equipment for the 410B.

"Actually," Doug continued, "we began cutting the first material for what became the **Model 350** POC (proof-of-concept) vehicle.

"A new nose section was built by Scaled Composites from available 324 tooling and delivered to us. We had an extra 324 horizontal tail and a set of elevons left over from the static structural test 324 airframe.

"Everything else we built from scratch — the whole center fuselage, the vertical tails and the wing. The wing was slightly shorter that the 324's so it was just as easy to build a new one.

"Though similar to the 324, the 350 POC has the bifurcated inlet instead of the single top inlet. Also, the fuselage is somewhat shorter and has hooks (bomb shackle fittings scrounged by Larry Emison) installed for air launch.

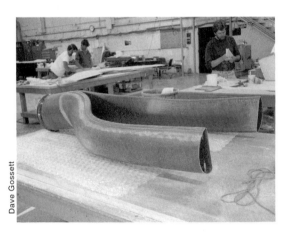

BIFURCATED INLET SYSTEM for jet power plant *is shown above. Below, the 410B fuselage takes shape.*

"They were doing the engineering and proposal effort down at one end of the plant while we were building the so-called 410B (actually the 350 POC) at the other. We took the three-view drawing on the day Mitchell said 'go' and went from there on our own.

"Understandably, Bob Mitchell wanted to keep our operation a closely-held secret from the competition so they couldn't counter it. To make that possible we didn't tell anyone in the company unless they had a 'need to know.' Most of the proposal people had no idea we were building a POC of the plane.

FAST PROTOTYPING TEAM completed the 410B vehicle *in less than two months in a secret 'skunk works.' Project head Doug Fronius, far right. ('Varieze' Model 400 - see page 152 - sulks in the far corner, left.)*

"On the 58th day, we had a roll-out party inside Building 183. By then, the bird was painted and ready for systems installation. The entire proposal team was invited to see what we, the Fast Prototyping team, had been doing. When they walked in they expressed amazement that we were able to build a 'mock-up' so fast. Mock-up? Well, no.

"When we told them it would be on the wing of an F-4 in two weeks, ready for captive flight, they thought we were crazy."

The POC bird was promptly shipped off to TRA's Mojave facility where Norman Sakamoto's experienced team installed all the internal systems, most of which had already been developed for the 324.

ROLL OUT OF THE '410B', now called the Model 350 POC proof-of-concept vehicle. *At left, the new UAV has bifurcated forward fuselage inlets, whereas the Model 324 Egyptian Scarab, right, sports a single top inlet toward the rear of its fuselage.*

Dave Gossett

Larry Mesecher from the Model 324 Flight Test Team was drafted to redesign the 324 electrical system including a prelaunch control box to allow air launch from an F-4C. Sherm (Lenny) Linen and Lloyd Morrison fabricated the modified hardware which was integrated into the airframe produced by Doug Fronius' team.

All of this effort took only five weeks. Tom Riley and Bob Coon, as a part-time effort from proposal activities, modified the airborne software in three weeks to accommodate air launch.

The timing could hardly have been better. For months, engineers and technical specialists had been working round-the-clock converting accumulated RPV design knowledge into a convincing proposal. The summary was short and succinct, but the complete proposal package was voluminous. It was shipped off to the potential customer at the end of August just as the 350 POC was beginning flight tests.

THE 350 POC ARRIVES AT MOJAVE to begin the test program.

MATING THE MODEL 350 TEST VEHICLE to the F-4 launch plane.

Dave Gossett

Dave Gossett

179

PROOF-OF-CONCEPT

FIRST CAPTIVE FLIGHT was completed Sunday, August 28 at Mojave with the 350 proof-of-concept vehicle on the wing of an F-4C operated by Flight Systems, Inc.

While there would be no problem of ground clearance with the 350 carried under wings of a four-engine DC-130, when the same vehicle was mounted under the F-4C's wing, clearance was a matter of inches.

The flight test vehicle, slightly smaller than the 324, is just over 16 feet in length and has an 11-foot wingspan. Like the Model 324, it has a twin-tail arrangement. Power is provided by a Teledyne CAE 373-8C turbojet engine delivering 970 pounds thrust. Maximum speed is Mach 0.9 and cruise range is 900 nautical miles.

Construction was of fiberglass and epoxy composites with graphite/epoxy composites in high-stress areas. This is the type of light yet strong construction favored for unmanned aircraft by many in the military.

AMPLE CLEARANCE HAVING BEEN VARIFIED between the bird suspended beneath the jet fighter *and the runway, the F-4 launch plane is ready for its take-off clearance from the Mojave tower.*

One nagging problem during takeoff on captive test flights concerned electro-magnetic interference which had to be shielded from the on board computer system.

The key to finding a successful solution was correctly identified, and ordered carried out, by the company's on-site Pilot-Engineer-President, Bob Mitchell, whose own reputation as a 'can do' hands-on executive was also at stake.

TRA's PR man, Jack Broward, saw to it that in-flight photos of the 350 snugly tucked under the wing of the F-4C promptly appeared in the trade press.

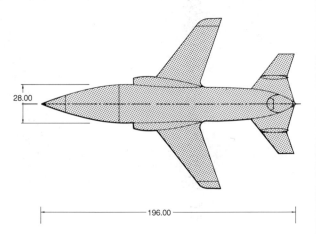

During the next few weeks, additional captive flights were flown to work out interface problems between the launch fighter and the proof-of-concept 350. All captive flights were cautiously flown to check out the basic propulsion system, power unit and guidance and navigational systems.

Not until every aspect of performance which could be checked in captive mode was proven satisfactory was the bird certified for free flight.

All risks at a minimum, a 'textbook perfect' first free flight was conducted at Mojave on Saturday, October 22, 1988.

"IT WAS AS SMOOTH and perfect as any flight test I have ever seen," remarked Flight Systems' John Ligon who had observed and photographed the 29-minute maiden test from an accompanying F-100 chase plane.

"The first flight went precisely as planned. The bird came off smoothly, climbed out and performed all programmed maneuvers according to our test objectives."

Veteran TRA drone controller Paul Bunner, manning the flight control console alongside the airport's main runway, found Ligon's observations particularly gratifying.

A 34-year TRA employee, who since 1965 has been associated with remote control operations, recalls that "all commands were crisp. I could hardly determine from my instrument readouts that the launch had been executed ...the vehicle provided no attitude disturbances or characteristics indicating discrepancies of any kind."

Bunner said the most critical area of concern was directed toward the vehicle's separation from the launch fighter aircraft.

Air-to-air photos by JOHN LIGON

"We launched at 15,000 feet and went into an immediate series of turns, banks and climb maneuvers, eventually to 27,500 feet." Recovery after flight maneuvers was made by parachute.

Sakamoto, who had fine-tuned the entire operation of the 324 and 350 at Mojave, expressed the view that the first flight "may well have exemplified the very best qualities of engineering design and development of which we are capable."

Not unexpectedly photos of the first free flight of the 350 POC found their way into the aviation press. If other firms in the MR-UAV competition were surprised by the August 28 first *captive* flight, they must have been startled by the initial *free flight* seven weeks later.

Some may have thought the captive flight involved a 'mock-up' rather that a flight-qualified vehicle. That idea evaporated with the 'near perfect' October 22 free flight.

Meanwhile proposals of all competitors were being closely scrutinized by the Navy and Joint Project Office.

Although a demonstrated flight vehicle was not required as part of the proposal, TRA had surprised everyone with the swiftness of its response and commitment to invest heavily in the now flight-proven 350 test vehicle.

Dave Gossett

OCTOBER 22, 1988. ALL'S WELL that ends well, with a 'good chute' recovery!

AND THE WINNER IS. . .

AWARD OF THE MR-UAV development program to TRA was announced May 5, 1989 by the Naval Air Systems Command. An interim $200,000 contract for engineering analysis accompanied the announcement.

Then, on June 30, NASC announced a $69.6 million award for full-scale development of the Mid-Range UAV through February 1992 and participation in integration of the new system into the Navy, Air Force and Marine Corps force structure.

Bob Mitchell's gamble, it seemed, had paid off.

Now it would be up to Norman Sakamoto, 32-year TRA veteran Program Manager, and his dedicated associates, to bring the project to fruition.

Three months after award, the initial design review conducted by the Navy went well. Sakamoto was pleased with the results and commented, "My belief is that the four-day period of review helped clear the air in many areas of program activity. The benefits that it provided us also apply to our customer.

"We explored many facets of understanding not previously discussed. In my years of responsibility for new programs, I've learned that it takes several face-to-face meetings to fine tune customer interfaces to maximum performance levels."

The more formal preliminary design review (PDR) conducted by NAVAIR in December '89 indicated continued satisfaction with the 350 program. At the time, one source commented

NAVY CAPTAIN JOHN OLMSTEAD (cut in half by UAV's fin) is briefed on BQM-145A details by TRA's Bob Mitchell during design review.

that "The contractor is building what we specified and he is on schedule."

What the contractor had developed and the customer contracted for was a vehicle constructed of composite materials. Somehow the materials experts at NAVAIR unexpectedly and belatedly raised questions of structural integrity of the airframe.

The question in the minds of those responsible for flight certification of Navy aircraft was whether or not possible delamination of graphite/epoxy composite surfaces might compromise vehicle integrity, especially during launch. For example, would the launch plane — and its pilot — be at risk?

Would some of the advantages of low radar cross-section (RCS) be sacrificed if metal replaced composites?

Should reusable unmanned vehicles meet the same maintainability standards as manned aircraft?

For many months in 1990, the future of the MR-UAV program was cloudy as contractor and customer tried to find a mutually acceptable solution.

Preliminary Design Review, December 20, 1989
"THE CONTRACTOR IS BUILDING WHAT WE SPECIFIED AND HE IS ON SCHEDULE.."

FOR NAVY AND MARINE CORPS OPERATIONS, the launch aircraft *will be the F/A-18. Two BQM-34A UAVs are shown tucked under the Hornet's wings in artist's concept.*

Meantime, criticisms ballooned around management of the program. The House Armed Services Committee, for one, got an oar into the troubled waters, pointing out that "the technical issues should have been identified by the Navy during the acquisition process before contract award; not two years later."

The best suggestion to solve the dilemma was in the form of a question.

Why not complete a few more composite vehicles; then, if study suggests the need for design and construction changes, switch to a metal structure for the production run? That would permit flight tests to continue with the Model 350-type composite birds as a risk reduction measure to validate or refine aerodynamic characteristics and subsystem performance.

Any production vehicles which followed would have the new official nomenclature **BQM-145A**.

RECOVERY OF UAVs at the conclusion of their missions is a major consideration in the Mid-Range program. Such has been the contention of OpNav flag officers.

Because of corrosion and decontamination problems associated with 'wet' recoveries at sea, the Navy has asked that special attention be given to providing the best technique to assure 'dry' recoveries.

The original MARS Mid-Air Retrieval System was pioneered by TRA for wartime operations in Southeast Asia during the late '60s. A similar system was designed for probable use in the 350 vehicles.

Once 'snatched' in mid-air, the bird can then be delivered 'dry' to an aircraft carrier, to other ships at sea or to a land base.

Under a Navy-sponsored research contract, TRA developed and demonstrated in summer 1990, a steerable parafoil mid-air retrieval system to determine whether it might offer significant advantages over the round parachute MARS equipment.

Also demonstrated was the ability of the Navy SH-60 Seahawk helicopter to perform its retrieval role.

TRA had contracted with Vertigo, Inc., to construct the tandem wing-type parafoil for the test program.

Inherently stable, the parafoil provides a consistent forward velocity that can simplify in-flight snatches by the retrieval helicopter. For the tests, a steering capability was added, allowing control of the flight path, on non-MARS missions, to soft landing of the vehicle.

Avtel Flight Test of Mojave provided the F-4D launch plane and test management support; the Navy furnished the Seahawk heli-

183

**DURING EARLY PARAFOIL TESTS, a cylindrical 'bomb'
simulated weight** *of the drone. Helicopter then snatches upper
foil to start mid-air recovery.*

nylon tubes rather than an umbrella, the parafoil has proven to be extremely maneuverable and can be landed with feather-light precision. The chore of the combined design and test groups was to adapt this type of parachute to new UAVs under development.

The MARS retrieval system (with capability to use either round or parafoil canopies) is provided for the BQM-145A under a Supplemental Agreement. Direct descent parachute recovery to the ground or sea is, of course, also available.

PARAFOIL PACKAGE IS EXTRACTED from top of proof-of-concept UAV *during instrumented evaluation test.*

copter and crews; Edwards AFB supported the Mojave-based tests with range facilities and loan of a MARS practice system.

After initial drops and retrievals with a cylindrical test vehicle, the complete system was flown on August 29 on the 10th evaluation flight with an instrumented 350 POC vehicle serving as a representative BQM-145A.

The 'BQM' was able to telemeter parachute and vehicle data to a ground station during flight, providing altitude, speed and other information. Remote controller steering had proven to be an asset in precision land recoveries during initial dummy drops. With a pre-landing 'flare' maneuver, just before touchdown it is indicated that impact damage should be negligible.

Anyone witnessing precision manned parachute drops at air shows around the country has become familiar with the parafoil parachute. Looking like a horizontal collection of

POTENTIAL PROBLEMS, in addition to considerations of structure and of 'dry' retrieval, continued to stalk what was subsequently termed the UAV-MR program.

The only mode of Navy and Marine Corps operation of the BQM-145A would be air launch from tactical aircraft including the versatile F/A-18, capable of aircraft carrier takeoffs carrying two recce birds.

A paramount consideration had to be flight and crew safety. Assurance was essential that at air launch, and during the post-launch phase, the UAV not accidentally fly into the launching platform.

Also at issue was the case where a UAV might have to be jettisoned during carrier catapult launch of the F/A-18, or during high angles of attack and varying multiple UAV and stores configurations.

During extensive scale model wind tunnel tests conducted with McDonnell-Douglas during the spring of 1991, all aspects of the areas of investigation were concluded satisfactorily.

AERODYNAMIC COMPATIBILITY of tactical F/A-18 launch aircraft *and its two BQM-145A UAVs was assured in scale model tests in wind tunnel.*

PROGRAM REDEFINITION

AFTER MONTHS OF DEBATE, extensive technical studies, possible reprogramming, specification and scheduling changes —brought about initially by the composites vs. metal controversy — the prospect of a solution began to surface.

While the technical discussions were under way, there was also an exchange of views going on regarding contract conditions, quantities, delivery dates, pricing and funding issues. For TRA, Waver Campbell, business manager reporting to Norm Sakamoto, carried the responsibility for these critical negotiations.

"The real problem," Waver points out, "was how to bring about a meeting of the minds where both parties could agree on a realistic program, within customer funding constraints, which would maximize benefits for both the user and the builder of the new RPVs."

After consideration of the many plans proposed, a Supplemental Agreement to the original contract between TRA and the UAV Joint Project Office was reached and formally announced June 13, 1991.

There would be a complete redefinition of the program, now known as UAV-MR, under which TRA was awarded an $11 million increment of funds under the previously awarded ceiling-priced contract of 1989.

Included in the new contract provisions was an increase in the ceiling price to over $180 million with increased flight testing and deliveries extending through early 1996.

TRA would build the first three vehicles of composite materials, and they would be used for flight test. These would be followed by 25 metal production BQM-145A mission-capable UAVs to complete the full scale engineering development (FSED) program.

The Mid-Air Retrieval System (MARS) as well as the direct parachute descent system would be available. Also incorporated in the final design are interchangeable nose configurations to match mission requirements as well as advanced technologies in guidance and control, navigation and data transmission.

Even before the supplemental agreement was effective, the first composite airframe, referred to as 'inert' IUAV, was completed. It would be used for static testing, engine installation, payload pallet fit, fuel tank pressure checks, recovery system validation and similar engineering tasks.

Two additional composite UAVs are to be used in the initial powered test flights.

185

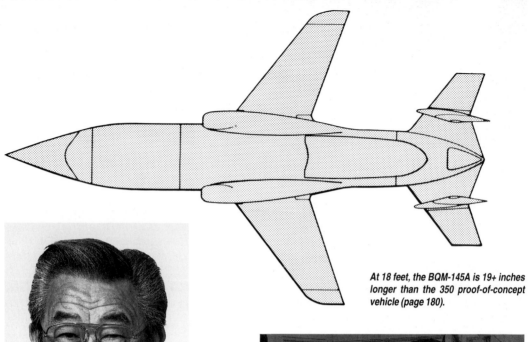

At 18 feet, the BQM-145A is 19+ inches longer than the 350 proof-of-concept vehicle (page 180).

Dave Gossett

VICE-PRESIDENT NORMAN SAKAMOTO who led the successful 324 Egyptian RPV Program, *has project responsibility for the BQM-145A contract.*

Dave Gossett

ROBERT A.K. MITCHELL, President of TRA, and REAR ADMIRAL GEORGE F.A. WAGNER, head the industry-military cooperative effort on the new UAV.

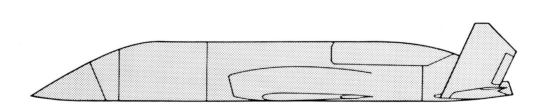

186

THE NO. 1 YBQM-145A composite-structure UAV *gets final inspection before formal roll-out.*

Dave Gossett

187

Joe Janes

KEY PROJECT EXECUTIVES include, above, Hudson B. Drake, *Teledyne, Inc. Senior Vice President and TRA President Robert A.K. Mitchell, right. Below, left to right, Navy Capt. John Olmstead, Army Col. Brad Brown and Marine Col. Larry Karch, all of the Joint Project Office.*

Dave Gossett

Joe Janes

JOINT PROJECT OFFICIALS, right, and TRA executives *gather before outdoor mural and stage in employees' plaza at roll-out of first YBQM-145A. Speaker is Capt. John Olmstead, Navy Program Manager. Company crew and roll-out vehicle, upper left.*

Joe Janes

THREE MONTHS AFTER the Supplemental Agreement was announced, Teledyne Ryan rolled out the first of the three YBQM-145 composite-structure UAVs, with representatives of the Navy, Marine Corps and Army present.

Heading the Joint Project Office attendees was Capt. John Olmstead, Navy program manager for the UAV-MR project.

"I've got the keys right here," Olmstead said, "Am I the first pilot?," adding that "all of the past concerns about the program have been addressed and it is now moving full speed ahead.

"Experiences of Desert Storm confirmed that we certainly need this capability."

In mid-December 1991, this same vehicle was released from an F-4 fighter at 12,000 feet over the Mojave desert and descended in an unpowered glide to 5,000 feet where its parachute recovery system was activated to lower the re-usable inert test YBQM-145 vehicle gently to ground contact.

"Our test objectives," said Paul Bunner, flight test director, "were to verify safe separation from the F-4 tactical aircraft, demonstrate total recovery system function, and assess any recovery or landing damage.

"We had a very clean separation from the fighter, normal function of the recovery system and complete acquisition of data. All test objectives were achieved."

UNDER THE FINAL CONTRACT redefinition provisions, TRA will have responsibility for integrating the principal intelligence-gathering payload into the BQM-145A.

Known as the Advanced Tactical Airborne Reconnaissance System, ATARS will provide optical or infrared target imagery on day or night high threat missions in heavily defended target areas.

During development of the ATARS equipment by contractors other than TRA, there were significant changes in that system's size, weight and other characteristics which require a somewhat longer airframe than provided for in the initial Model 350 proof-of-concept vehicle on which the BQM-145A design is based.

Operationally, on its return leg to base, the ATARS-equipped UAV can transmit recorded data to an image processing ground station. This is a capability of great value in tactical situations when battlefield commanders are vitally in need of real-time reconnaissance data.

DESERT STORM OPERATIONS 1990-91 against Iraq's vaunted military machine also offered lessons-to-be-learned in providing theater commanders with timely information including enemy capability.

The aviation press reported that current tactical intelligence, including battle damage assessment damage, often was not readily available in the fast-paced high-technology war. Only light, short-range propeller-driven UAVs were operational.

Although neither military or company sources have commented, it is widely believed that if the war against Iraq had continued there would have been a request made to Egypt to permit operation of its Model 324 reconnaissance Scarab RPVs in the Iraq theater.

TRA's President, Bob Mitchell, was quoted in the press as saying, "Given time, money and priorities, we could have solved the battle damage assessment problems."

PERHAPS THE DESERT STORM experience reinforces the concept that the United States should always have a quick-reaction, standby, unmanned reconnaissance/surveillance capability. A re-reading of "What Now?," on page 108, is pertinent.

───────────

PREPARATIONS GOT UNDER WAY early in 1992 to activate the full-blown flight test program at the remote Air Force and Army Test and Training Ranges southwest of Salt Lake City, Utah.

Equipment and TRA technical personnel, in March and April, began moving into facilities at Michael Army Air Field on the Dugway Proving Ground, 35 miles as the crow flies from the small town of Tooele. Operational control for the test program would be centered at Hill Air Force Base near Ogden.

As in earlier flight tests with the Model 350 proof-of-concept vehicle, captive flights of the YBQM-145A would precede release of the new composite-structure UAV for its initial powered flight. Again, the launch aircraft would be the F-4 Phantom operated by Avtel Flight Test of Mojave.

While attached to the F-4, captive flights at Dugway were conducted April 12 and May 1 as reliability checks and as a dress rehearsal for the upcoming first powered free flight.

FIRST FLIGHT: YBQM-145A

FOR AN ON-SITE ACCOUNT of the important first powered free flight of the YBQM-145A Unmanned Aerial Vehicle (UAV-MR), we asked Mark Day of TRA marketing communications to help out. Here is his report.

THE NEW 'BIRD' ISN'T LARGE...just 18 feet long with a wing span of 11.5 feet. But on May 5, 1992, the compact YBQM was the biggest thing going on at the Utah Test and Training Range that covers thousands of square miles of arid desert west of Salt Lake City.

Precisely at 12:00 noon, a smokey-grey F-4 Phantom jet fighter taxied onto the 11,000 foot runway of Michael Air Facility at the U.S. Army's Dugway Proving Ground.

The nation's newest, high-performance unmanned surveillance aircraft, painted a bright international orange, was attached to the center pylon under the fuselage of the F-4.

SNUGLY TUCKED BENEATH THE F-4 JET, the YBQM-145A is ready to fly *as the launch plane taxis out for take-off accompanied by the F-16 chase plane.*

"....two aircraft, flying as one...."

From the opposite end of the Michael runway, the fighter could not be seen through the haze and glimmering heat waves rising from the desert floor, but as "Olds-1", the civilian test pilot, advanced his throttles, the pounding thunder of the F-4's engines soon became deafening.

"Olds-1" released his brakes and the F-4 began to build speed for a long, shallow take-off. Then, just as its nose came out of a desert mirage about halfway down the runway, the F-4 rotated up and the sharp profile of the UAV underneath became clear.

The two aircraft, flying as one, soon passed overhead on their way to 15,000 feet. The first free-flight of YBQM-145A...under its own power...was about to begin.

Cliff Pryor, one of TRA's flight test engineers on the tarmac that morning, as the UAV was uploaded onto the F-4, remarked that he has been "...living with the bird" for months. "Everyone up here has been working long hours, seven days a week," he said. "It has been tough, but they are all good people working on a good airplane. This flight will prove it."

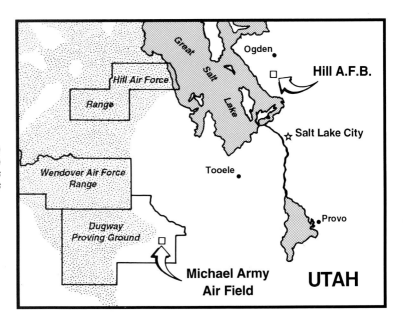

EXTENSIVE UTAH TEST AND TRAINING RANGES were the locale of new UAV's first free flight, *as they have been for other key tests staged from Hill Air Force Base.*

MONITORING FIRST FLIGHT were Gene Motter, TRA manager of test engineering, *and chief aerodynamicist Arnie Otchin. Below, Bert Hansen at Local Control Monitoring Station (LCMS).*

Those long hours at the remote desert test site began a relatively few weeks after the UAV had a successful unpowered free flight in December, 1991, to test both its ability to be launched safely from a tactical aircraft and its ability to be recovered by parachute.

At 12:10 P.M. MDT, "ABLE-1" (Fred Dukes and Bob Coon) the mission's test controllers at Hill AFB, announced "Launch minus 5 minutes." The F-4's pilot, ground controllers, the range safety officer, and chase plane pilot began the countdown.

After numerous system checks, telemetry checks and camera checks, the radio network again came to life with "All stations confirm readiness to proceed to engine start." At "Launch minus 3 minutes," the UAV's small but powerful TCAE 382-10C jet turbine engine was switched on.

In the TRA telemetry trailer on the ground at Dugway, company president Robert A. K. Mitchell intently watched the needles of the telemetry readouts that indicated engine RPMs. For 30 seconds they indicated just an engine idle. Ten more seconds...Still not fully up to power. Then the UAV's engine kicked up to full power and held. Mitchell, along with a dozen others in the small trailer cheered. "This is Able-1...engine ready indicator steady now, we have a green board!"

Almost precisely at 12:15 p.m., while everyone held their breath, the radio network sparked with "...minus five, four, three, two, one, launch," and Olds-1 announced from the F-4 "Air vehicle away, climbing up and left...we're clear!"

191

THE YBQM-145A WAS SUCCESSFULLY launched on an important free flight test that would last 37 minutes. From its launch at 15,000 feet at Mach 0.7, the bird flew both autonomously and under remote control from nearby Hill Air Force Base.

Throughout the flight an Air Force F-16 chase plane kept the UAV in sight, sometimes closing to within 80 feet of the glistening orange aircraft. The photos and video tape taken from the back seat of the F-16 by Staff Sergeant Martin Hamilton, USAF, proved to be as spectacular as the flight

The UAV climbed up to 30,000 feet, maneuvering at speeds up to Mach 0.8 with the snow-covered caps of the Utah mountains below mixing with the few clouds that were drifting across the range that day.

After cruising at 30,000 feet and following a predetermined test flight pattern around the test and training range, the UAV was instructed to descend. At 20,000 feet Able-1 announced on the radio network, "Test Conductor initiating recover on my mark. Three, two, one, Mark."

Instantly, a hatch cover on the top of the UAV's fuselage blew off and a small drougue chute popped out, pulling a second medium-sized chute out of the airplane that in turn deployed the UAV's large recovery parachute.

Air speed began to drop as the chute fluttered and then mushroomed into shape as it

grabbed air. The bird's critical recovery system performed perfectly, and the YBQM-145A began a slow, gentle descent to the Utah desert floor. Minutes later, a HU-1 Huey recovery helicopter with TRA and test range personnel landed nearby. The bird was intact with no significant damage. It would fly again.

Sherman "Lenny" Linen, a TRA veteran who has had years of flight test experience with Remotely Piloted Vehicles and UAVs, could now relax, along with the dozens of other Navy, Air Force, TRA and civilian personnel who worked on the flight.

"All the experience we have in UAVs really helped us get ready for this one," Lenny said while inspecting the UAV. "We put lots of hours into prepping for today's flight and I feel good. All our experience has paid off...with success."

Dave Gossett

Dave Gossett

TECHNICIANS EXAMINE YBQM-145A after relatively soft landing on first flight. *HU-1 Huey jet helicopter from Hill AFB airlifts UAV back to Michael Army Air Field to fly another day.*

AS THIS "FIREFLIES" HISTORICAL book goes to press, there is still an extensive full scale engineering development (FSED) program ahead for the BQM-145A.

With 'first flight' on May 5, 1992 now a matter of record, the next phase will also involve demonstration of the UAV's ground launch capability.

Production of 23 metal-structure UAVs will follow the three composite-structure 'Y' birds which will continue the initial engineering test flights.

Current contracts call for deliveries extending into 1996.

Ssgt. MARTIN E. HAMILTON-U.S. Air Force

FLIGHT TESTS IN HIGH GEAR!

IN A TIGHTLY-PACKED 12-DAY PERIOD July 25 to August 5, 1992, four different models of TRA unmanned aerial vehicles – 410, 350, 324 and the newest, the 'platform' for the 'ARGUS' electro-optical payload – conducted significant first flight tests and proving missions.

Two projects – the air-launched BQM-145A mid-range UAV and the Argus 'platform' flights – were staged from Hill Air Force Base in Utah. First flight of Argus was made August 5th.

Dave Gossett

Captive beneath the wing of its four-engine launch plane, the UAV 'platform' for the Argus electro-optical payload was later released for a successful first free flight on its own power August 5, 1992.

Assigned to the four UAV models were four matching teams of technical experts, each with its own program manager and support staff.

The runway-based, piston-engined 'affordable' Model 410 operated from Cuddeback dry lake on the Edwards Air Force range in California, making its first unmanned runway take-off, flight and landing July 25th.

Three 324 Scarab RPVs were flown in-country in five days, July 28 and 29 and August 1, by the Egyptian Air Force, using their unique land-based mobile launch and recovery vehicles (LRVs).

As a demonstration of versatile operating capability with multi-design vehicles, the July-August fortnight was a record-setting, stellar performance in pioneer Teledyne Ryan's 70-year history.

THE MAKING OF A BOOK

Authors need plenty of help in getting a new book off the press and into the readers' hands.

The production cycle of the "FIREFLIES" book started with word processing; in this case by Linda Sprekelmeyer. Graphic artists Randy Crawford and Kent Rump mated the type with crisp photos printed by Joe Janes.

After final review by Publisher Jay Miller, the complete package was shipped off to the Printer for production.

A special word about the illustrations.

TRA's veteran chief photographer, Dave Gossett, took most of the photos.

His work was supplemented using outstanding chapter-heading photos by such camera artists as Chad Slattery (page 75); the Air Force's James P. Porter (109); TSgt. Michael James Haggerty (148) and Ssgt. Martin E. Hamilton (175); John Ligon (181), and TRA's on-site technicians — Sherman Linen in Egypt (157) and Steve Uyehara in Japan (65).

As to the narrative text it would be colorless but for the in-depth interviews with technicians, field service representatives and program managers.

Not all can be mentioned but we would be negligent not to credit Doug Fronius, Larry Emison, Billy Sved, Paul Bunner, Frank Marshall, Bud Miller, Frank Oldfield, Bob Perkins, Mike Savino, Jack Young, Dick Manceau, and on and on.....

And, yes, the proofreader-indexer-grammarians — Audrey Ireland and Sally Ades — who tried to keep the authors from wandering too far astray.

It is also appropriate to mention the co-operation of the foreign press, particularly the Israel Air Force Magazine, the Jerusalem Post and Israel Aviation & Space Magazine.

Ned Porter of the Bangor, Maine News also provided an excellent account of his observations while aboard an Air Force DC-130 as it launched an RPV on a special mission off the Atlantic coast.

INDEX

Page numbers in **bold** type indicate major references.

A

AQM-34 (Model 147) series
  Reconnaissance/ECM vehicles **1-13**
AQM-34L/M (Model 147SC/SD) 142-150
AQM-34V (Model 255) Combat Angel
  ECM .102-104
AQM-81A/N (Model 305) Firebolt 124-129
AQM-91A (Model 154) Compass Arrow
  (Firefly) **27-47**
Also see more detailed listing under 'Models'
  Note: A = air launch
              Q = drone
              M = missile
A-4 Skyhawk 11,53,127,129
A-6 Attack aircraft 131
A-10 Attack aircraft 117
Abdallah, A. El Zohiry, General, Egyptian
  Air Force 173
Abu Ghazala, Field Marshal Mohammed
  Abdel Halim 159-163
Advanced Tactical Airborne Reconnaissance
  System (ATARS) 189
Aegis System aboard Cruisers **148-150**;
              also see 90,93,96,120,142-145
Aeronautica Macchi (Aermacchi) 72-74
Aerospace Daily 31-32
AIM 4-F Missile 25
AIR FORCE . . .*also see Strategic Air Command*
              and Tactical Air Command
  Aeronautical Systems Division (ASD) . . . 105,
                        106,110,111,116
  Air Defense Command (ADC) 102
  Air Defense Weapons Center, Tyndall
    Air Force Base, Florida . . . . . .*see Tyndall*
  Air Material Command (AMC) 15,146
  Armament Development Center, Eglin
    Air Force Base, Florida 126
  Electronic Systems Divisions 145
  Legislative Liaison 32
  Logistics Command (AFLC) 6
  Pilotless Aircraft Branch 15
  Reconnaissance Division 1-2
  Systems Command (AFSC) 39,98,142
Air National Guard, Maine 145
Airfield Damage Assessment System
  (ADAS) 136-137
Alamaze Air Base, Egypt 163
Alamogordo, New Mexico 34,42
Alaska 145-146
Albany, USS 74
Albuquerque Journal 29,34
Alexandria, Egypt 62
All American, Inc. 136
Allen, Gordon 57
Alouette Helicopter 59
America, USS (CVA-66) 79,90,148
Ames Research Center (NASA) 121
Amori Prefecture, Japan 67
Anagada Island, West Indies 25
Anderson, Senator Clinton P. 30
Anderson, Maj. Rudolph, Jr. 1

Andrews Air Force Base . . . .
Anti-Ship Missile Target (ASMT)
  (Firebrand) **119-124**
Anti-Submarine Rocket (ASROC) booster 96
Apache (AH-64A) helicopter 93,160
'Apple Splitter' Trophy 81,83
Arab Republic of Egypt **157-174**
Arctic Ocean 146
Ark Royal, HMS (British Carrier) 25
Armstrong, Neil 97
ARMY
  Army Missile Command (MICOM)
    Huntsville, Alabama 86-87,96
  Army Signal Corps 15,80
AS-3 Kangaroo Projectile 120
AS-15 Soviet cruise missile 143
Asher, Maier 48
Ashraf, Col. A. Khalil, Egyptian Air Force . . . 173
Associated Press 29,32
Association for Unmanned Vehicle Systems
  (AUVS) 18
Athena Program 32
Atlantic Fleet Weapons Range (AFWR),
  Puerto Rico 23-25,78,**88-90**,97,116,120
              also see Roosevelt Roads
Atomic Energy Commission 30,33-34
Auerbach, Cdr. Eugene E. 120-122
Aviation/Space Writers 111
Aviation Week & Space Technology 30,132
Avtel Flight Test 183, 189
Azon, Razon, Tarzon Glide Bombs 15
'Azuma' III 65-71,97

B

BQM-34 series (Firebee I) Training
  Targets **75-93**
BQM-34E/F (Model 166) Firebee II
  Supersonic Training Targets **19-26**
BQM-145A (Model 350) . . . . **175—193**
BQM/SSM (Model 248) Surface-to-
  Surface Weapon Delivery **96-98**
ZBQM-111A (Model 258) Firebrand . . **119-123**
BGM-34A/B (Model 234) Multi-Mission . **96-101**
BGM-34C (Model 259) Multi-Mission . **105-108**
Also see more detailed listings under 'Models'
  Note: B = air and/or ground launch
              Q = drone
              G = surface attack
              M = missile
              Z = planning
B-1 bomber 112-113
B-47 bomber 9
B-52 bomber 10,86,108,123
'Baby Buck' 6
Babcock, Col. Earl 17
Back-up Director System (BUDS) 55
Bagwell, Charles B. 17
Ballweg, Raymond D. 53,58
Bangor, Maine 143-150
Bar-Lev, Lt. General Haim 50,54,60

Barber's Point (Barking Sands), Hawaii . . 89,150
Barometric Low Altitude Control System
    (BLACS) **4**,96
Barter Island, Beaufort Sea, Alaska . . . 146-147
Beckner, Major Ken 33
Beech/Martin Marietta 176
Bell, Rep. Alonzo 32
Bell 'Huey' helicopter 66
Bendix 120
Bermuda 143-144
Berner, Carroll 21,79-80
Berry, Bill 73
Bien Hoa Air Base, South Vietnam . . .2,4,6,8,142
'Big Red' (BGM-34C) 95
'Big Safari' Program 1,2,7,21,36,
                            39-40,102
Blackwell, H. Jack 30
Blanchard, Maj. Gen. William H. (Butch) . . . 102
'Blue Book' 36
Boeing Aerospace 110-118
Bomb Damage Assessment (BDA) 108
Bonneville Racer 36
Boone, Robert T. 17
Bowes, Rear Adm. William 18
Boyington, Pappy 78
BQM-106 research RPV 136
Brady, Jim 24
Brennan, Tom 133
Bright, Denise 155
British Navy 25,110
Broward, Jack 180
Brown, Col. Brad 188
Brown Field, San Diego 138
Brunswick Corporation 131
'Budweiser' RPV (Vietnam) 6
Built-In Test Equipment (BITE) . . .43
Bunganich, John 22
Bunker Hill, USS 148
Bunner, Paul 85-87,165,166,169,180
Burns, Darrel 30
Busy Robot 45
Butler, USS 98

C

C-5A Galaxy cargo plane 112,115,146
C-130 Hercules cargo plane 56,74,89,143,
               144,150,165; also see DC-130
CH-3E helicopter 38,43,58,103
Cairo, Egypt 50,60,62,158,161-164,170
Campbell, Dave 127
Campbell, Waver 185
Canada 145,146
Cape Canaveral, Florida 115,116
Captive Test Vehicles (CTV) 40
Carnes, Col. Fred 127
Carpenter, Paul 57
Cathey, Oliver 160-163
Central Intelligence Agency (CIA) 36,40
Chaff dispensing 9,**103-108**,150
Chaparral missile 84
Chemical Systems Division (United
    Technology) 126,131
China 2,3,7,9,15,35,36,45,47,66

China Lake, California . . . .see Naval Ordnance
                                 Test Station
Chinook helicopters 169,172
Christian, Ed 5
Clark Air Force Base, Philippines 88
Clark, USS 149
Clasen, William 36
Close-In Weapons System (CIWS) 120
Cohen, Col. 'Cheetah' 54-56
Combat Air Patrol (CAP) 5
'Combat Angel' 102-104
'Combat Dawn' missions 110,115
ComNavAirPac 24
Compass Arrow (AQM-91A) **27-47**;
             also see 110-112,160,161,164
Compass Cope-R (YQM-98A) 29,**109-117**,
                   132,135,155,160
Composite Squadron 3 (VC-3) 89
Composite structures 160-162
Continental Aviation and Engr (CAE)
    J69 and other jet engines . . . 4,6,9,16,21,37,
         52,70,86,87,126,160,166,180, 191
Continental Motors Corp. 37,51,158
Continental piston engines 134,162
Control and other TRA systems . see alphabetical
                listings for the following:
    BLACS (Barometric Low Altitude Control
      System)
    BUDS (Back-up Director System)
    HARS (Heading and Altitude Reference System)
    LRV (Launch and Recovery Vehicle)
    MARS (Mid-Air Retrieval System)
    MASTACS (Maneuverability Augmentation
      System for Tactical Air Combat Simulation)
    MFCS (Microprocessor Flight Control System)
    MLCU (Mission Logic Control Unit)
    RACE (Radar Altimeter Control Equipment)
    RALACS (Radar Altimeter Low Altitude
      Control System)
Control Data flight computer 121
Coolidge, President Calvin 80
Coon, Bob 179, 191
Copley News Service 110
Coronet Thor 101
Corra, Lt. Col. Andrew J. 6
Cote, Ray 155
Crete 50
Critical Design Review Board (Air Force) . . . 127
Cuba 1,2,46,114,115
Cubi Point Naval Air Station, Philippines . . . 79
Cunningham, Rep. Randall (Duke) 78
Curtiss Robin Monoplane 15
Customs Service (U.S.) 156
Cyclone Aviation Products Ltd. 50-53

D

DC-130 Hercules launch planes . . .2,3,5,8,22-26,
       30-33,38-45,76,80, 87-89,98-104,
       108,121,123,127,131,143-150,180
DP-2E Neptune launch planes 23,24,78,
                   79,89,105
Daily Express, Daily Telegraph 48
Dale, Col. John 4,6
DaNang, South Vietnam 3,4,6

Dan Hotel, Tel Aviv 56
'Dark Eagle' program 30
Davies, Ivor 48
Davis-Monthan Air Force Base, Arizona . . 45,47,
                                   103-104,110
Day, Mark 190
Dayan, Defense Minister Moshe 60
Dayton-Wright Company 47
Defense Advanced Research Projects
     Agency(DARPA) 135
Defense Department 3,18,49,98,
                           107-108,118,128,139
Demilitarized Zone (DMZ) 9,102,103
Denny, Reginald 15,135
Desert Storm, Iraq 87,148,189
Dickens, Billy 166
Digital Design and Manufacturing 136
Digital Flight Control System 71
Direct Control Operater (DCO) 5,45,89
Director of Defense Research and
     Engineering(DDR&E) 135
Distance Early Warning (DEW) radars 145
Doering, Joe 57
Donaldson, A.M. 79,80,98
Doppler Radar Navigation 43,45
Dotson, Gene 177
Dov Airport, Israel 56
Dowell, Bill G. 20,21,24,26
Drake, Hudson B. . . 93,132,133,152,163,164,176
Dror, Samuel 'Bondi' 51,52
Dugway Proving Ground, Utah 100,102,106,
                                      189, 190
Dukes, Fred 191
DuPont Kevlar 134,150,159,160
Durango, Colorado 42

E

EC-121, Super-Connie 9
11th Tactical Drone Squadron (TDS) . . . 103,104
Eagan, Brian 164
Eason, Capt. W.E. 91
Eastern Test Range, Patrick Air Force Base,
     Florida 114
Edwards Air Force Base (EAFB), California . . 14,
             25,31,44,45,99,100,111-115,177
Eglin Air Force Base and Gulf Test
     Range, Florida 103,127,128
Egypt, Arab Republic of 157-174;
          also see 48,50,54,60-63,96,120
Eisenhower, President Dwight D. 49
El Centro, California . . . . see Parachute
                                   Test Facility
Elath (Israeli Destroyer) 96,120
Electronic Countermeasures (ECM) and
     Electronic Warfare (EW) 22,38,46,95,
                           96,102-108,150
Elkhorn Flare System 90
Elmendorf Air Force Base, Alaska . . 83,146,147
Emison, Larry 35,43,45,67,68,
                          87,126,127,178
EPSCO Track and Control System 69
Erez, Oded 50
Evans, General William 115
Exocet missile 139,148

F

Firebee *-The generic term used since 1950 to describe a wide variety of Teledyne Ryan unmanned aerial vehicles.*
Firebee *-Early history -* **15-16**. *Covers models Q-2A (Air Force), XM-21 (Army), KDA-1 (Navy), and KDA-4 (Navy and Canada).*

Upgraded Model Q-2C has BQM-34A designation for Air Force and Navy, and MQM-34D for Army.

(See chapter on Pilotless Target Aircraft pp.75-93)

Firebee I (BQM-34 and MQM-34) Targets 75-93
Firebee II (BQM-34E/F) (Model 166) . . .**19-26**
Firebolt (AQM-81A/N) (Model 305) . . **124-129**
Firebrand (ZBQM-111A) (Model 258) . **119-123**
Fire Fly (BQM-34A) (Model 147A)
     'Lightning Bugs' **2,102**
Firefly (AQM-91A) (Model 154)
     Compass Arrow **27-47**
Also see more detailed listing under 'Models'

432nd Drone Generation Squadron
     (Tactical Drone Group) 100,103,104,108
4080th Strategic Reconnaissance Wing
     (SRW) 104
4750th Test Squadron 25
Falcon missile 26,81
Falklands War (1982) 92,110
Fansong Radars 46
Farmington, New Mexico 31,42
FDL-23 research vehicle 100,104
FDL/BQM-106 mini drone 136
Federal Aviation Administration (FAA) 32,44,114
Fekry, Maj. Elsayed Shanin, Egyptian
     Air Force 173
Fighter Aircraft
     F-4 Phantom Jet 14,25,26,51-55,59,78-80,
          83,91,98-100,108,131,177-181,183,189-192
     F-9F Cougar 91,116
     F-14 Tomcat 80
     F-15 Eagle 113
     F-16 Falcon 177, 190, 192
     F/A-18 Hornet 131,177,183-185
     F-86 Sabre 91
     F-100 Super Sabre 83,180
     F-101 Voodoo 26
     F-105 Thunder Chief 100
     F-106 Delta Dart 26
Firebee Low Altitude Ship-to-Ship Homing
     missile (FLASH) 96
Fitch, Capt. Robert 149
Flader XJ-55 jet engine 126
Flax, Dr. Alexander H. 37
'Flexbee' 135
Flight Dynamics Laboratory (Air Force) 100,
                                     104,136
Flight Systems, Inc. 180
Flying Boxcar 56
Forehand, Col. William W. 4,6,56
Fort Huachuca, Arizona 31,44
Forward Looking Infrared (FLIR) system . . . 154
Four Corners (Utah, Colorado, New Mexico,
     Arizona) 44
FR-1 Fireball Navy jet fighter 16,110
Franklin D. Roosevelt, aircraft carrier 25

Fronius, Doug 152,153,160,162,
165,177,178,179
Frosch, Dr. Robert A. 96
Fuji Heavy Industries 66-71

G

GB-1 and GB-4 Glide Bombs 15
'Gallant Eagle' exercises 103
'Gap Filler' 145,146
Garrett AT F-3 jet engines 111,115
General Accounting Office (GAO) 118
General Electric Co. 37,47,142
General Electric jet engines . . . 39,87,88,110,117
Geneva Conference (1955) 49
Germany 25,101
Ghazala, Field Marshal Mohammed
Abdel Halim 159-163
Giza, Egypt 158,169
Globe Position System (GPS)
navigation 154-156,166
Golan Heights 52
Goolsby, John 91
Gossett, Dave 147
Gould, Jack 49
Grady, Jeff 57
Green River, Utah 31
Grenard, W.A. 134,138,160-162
Ground Launch 84-85
Guidance Navigational System (GNS) 45
Gulf of Sidra 148
Gulf of Tonkin 2,5,46

H

H-3 helicopter 146,147
H-53 helicopter 56,62
HU-1 Huey helicopter 192, 193
Hafez M. Zaki, Egyptian Air Force 161,173
Haiphong, North Vietnam 5,6,10,11
Hall, Dave 163,164
Hamilton, Ssgt. Martin E. 175, 192
Hamilton, Walt 22,45,88
Hamrick, B.R. (Bob) 140,141,156
Hanoi Hilton 11,100
Hanoi, North Vietnam 6,10,11
Hansen, Bert 191
Hanscom Air Force Base, Maine 142,145
Harper, Dave 153
Harpoon jet engine (CAE) 160,166,180
Harpoon ship-to-ship missile 96,98,121
Harrier Jet Fighter 92,110
'Have Lemon' 98
Hawk missile 26,66,67,72,74,84,85
Heading and Altitude Reference
System (HARS) 43
Heil Ha'avir (Israel Air Force Journal) 61
Heilman, N.C. (Butch) 18
Heliopolis, Egypt 158,159
Helmich, William F., Jr. 98,100
Hemenway, Col. Ward W. . . . 100,105,110,112
Henninger, Harry 34
High Altitude High Speed Target
(HAHST) (Air Force) 123,126
High-Altitude, Long-Endurance (HALE) 116,
118,132,134

High Altitude Supersonic Target (HAST) . . . 126
High Altitude Target Skylite (HATS) . . . 140
High Energy Laser (HEL) Weapons 120
High Energy Laser System Test Facility,
White Sands, New Mexico 139
Hill Air Force Base, Utah 26,100,101,
106,142-150, 189-192
Hind D Mi-24 Soviet helicopter 138
Hirst, W.R. (Bill) 161,163,169
Holloman Air Force Base and Develop-
ment Center, New Mexico 6,14,29,32,
34,40-44
Holtville, Imperial Valley, California . . 154-156,
177
Honeywell 121
House Armed Services Committee 183
Hughes Aircraft 139-140
Huntsville, Alabama 86
Hurlburt Field, Florida . . . .
Hurley, Larry 166

I

Increased Maneuverability Kit (IMK) 78,79
Infrared (IR) 38,164
Initial Operational Test & Evaluation (IOT&E) 128
Instrument Landing System (ILS) 154
Integrated Track and Control System (ITCS) . 87
International Maritime Rules 144
Intripid, USS 11
Iran Air 148
Iran-Iraq War (1982-88) 139,148
Iraq 64,87,90,139,148,150,189
Israel 20,**48-64**,87,96,120,158
Israeli Aircraft Industries 51
Ita-Jima 68
Iwo Jima 67

J

Jackson, Captain 42
Jackson, Robert C. 11,37
Jameson, Frank Gard 24,52,58
Japan and its Self-Defense Forces **66-71**
Jerusalem 52
Jerusalem Post 48,59
Jet Assisted Take-Off (JATO) . . 14,54,55,67,69,81
Johnson, President Lyndon B. 2,102
Joint Project Office (JPO) for UAVs and
Cruise Missiles 18,132,161,185
Joint Services Common Airframe Multiple
Purpose System (JSCAMPS) 176
Jordan 50
Juberg, M.E. (Gene) 99,100,103,104,166

K

KDA-1, KDA-4 . . . . . . 16, also see Firebee
KA-80A camera 38,45
KA-93 camera 63
Kadena Air Base, Okinawa 2
Kaktovik, Alaska 146
Kaminski, Tony 57
Karch, Col. Larry 188
Kasahara, Akimichi 66

Kellar, Gerald R. 138
Kettering Aerial Torpedo ('Bug') 15,18
Kettering, Charles F. 15
Kevlar . . . . *See Dupont*
Khrushchev, Nikita 49
Kibbutz, Israel 56,60
Kirtland Air Force Base, New Mexico 34
Kissinger, Henry 47
Klein, David 52
Knache, Ted 18
Kom Oshien, Egypt 167,169,170
Korea (South and North) . . . 9,46,84,92,110,115
Kuwait 64,150
Kyoto, Japan 67

L

La Maddalina, Italy 74
Laroce, Marshall 92
Laser weapons tests **139-141**
Launch and Recovery Vehicle (LRV) . 160,165-171
Launch Control Operator (LCO) 33
Lear Siegler 17,105
Lebanon 63
LeBaron, Al 106
Lee, Maj. Linus L. (Buck) 6
Libya .90,148
Ligon, John 180,181
Lindbergh, Charles A. 2
Lindbergh Field, San Diego 112
'Linebacker II' 10,108
Linen, Sherman (Lenny) . . . 169-173,179, 192
'Litterbug' Project 9
Litton Industries 166
Local Control Monitoring Station (LCMS) . . 191
Lockheed launch aircraft 89
                    also see DC-130, DP-2E
'Lone Eagle' 37
Loran Navigation 104,107,108,156
Los Alamos, New Mexico 29-34,42
Lucast, J. E. (Jack) 2
'Lucy Lee' 1
Luke Air Force Base, Arizona 44
Lycoming piston engine 153

M

MQM-34D (Army Firebee) 51,**86,87**
   (the first M denotes mobile ground launch)
MQM-8G Vandal 120,123
Mackey, Capt. W.A. 24
Maged, Gen. Sayed Aly 158,161,163
Maizuri Heavy Industries 67
Mallios, William 17
Manceau, Richard F. 24,89,90
Maneuverability Augmentation System
   for Tactical Air Combat Simulation
   (MASTACS) 79,80
'Manta Ray' (Model 262) 135,136
Mao Tse-tung 46
Mares, Ernie 78
Marine Air Detachment (Pt. Mugu) 24
Marquardt ramjet engines 120,121
Marshall, Frank 54,57,58
Martin Marietta/Beech 176
Martin, Vice Adm. Edward H. 11

Massachusetts Institute of Technology (M.I.T.) . 6
Masumoto, General 68
Matacotta, Gen. Mario 72
Maverick AGM-65 missile 98-101
McCarter, R.S. (Steve) 161,163
McDonnell-Douglas wind tunnel 184,185
McGean, Chuck 113
McGill, C.E. (Chuck) 164,169,172
McGregor, Jim 149
McGregor Range, New Mexico 52,67,68,128
McMahon, Markly and Jean 30
McMillen, Dr. Brockway 37
McVicker, Bob 17
Mediterranean Sea 25,148
Meir, Prime Minister Golda 98
Mena House, Giza, Egypt 158,170
Mesecher, Larry 167,169,172,173,179
Michael Army Air Field, Utah 189, 190
Microprocessor Flight Control System
   (MFCS) 63,64,88,104,156
Microwave Command Guidance System
   (MCGS) 38,44,55,99,113,147
Mid-Air Retrieval System (MARS) . **4**,15,22,25,26,
                    31,38,**40**,42,45,56,61,
                    99,103,127,129,144,183,185
Middle East 20,46-51,60,64,98,148
Mid-Infrared Advanced Chemical Laser
   (MIRACL) 139-141
Midi-RPV 133,161
Mid-Range Unmanned Aerial Vehicle
   (MR-UAV) 132,155,161,169,**175-190**
MiG Fighters 4,5,9,51,59-63
Miller, Barry 30
Miller, C.D. (Bud) 68,69,85,87,91
Mirage IV 50
Miramar Naval Air Station, San Diego . *see Navy
                    Fighter Weapons School*
Mission Logic Control Unit (MLCU) 166
Mitchell, Robert A.K. (Bob) . . . 155,172,**176**,177,
   178,182,186-191
Mobile Ground Launchers 84,85,165
Models . . . . **see next page**
Mojave, California operations 162,166,167,
                    177,178,180,181
Mokhtar, Gen. M. El Nomrossy, Egyptian
   Air Force 173
Morino, Maki 66,67
Morrison, Lloyd 169,173,179
Moser, Ernie 68,98
Motter, E.R. (Gene) 2,147,154, 191
Mt. Fuji, Japan 66
Mubarak, Egyptian President Hosni 163
Muklebust, Marvin 149
Multi-Mission RPVs **95-108**

N

96th Fighter Squadron 78
99th Stratigic Reconnaissance Squadron . . . 6
N9J Navy Aeroplane 15
NC-130 Hercules 143,144,148
Nadim, Gen. El Sayed Aly 158-163
NASA 144
National Aeronautics Association 127

continued on page 202

MODEL DIRECTORY

Unmanned Aerial Vehicles (UAVs) are available in all sizes, configurations and capability to meet a wide variety of customer requirements.

Here, to scale, are five of the three dozen models listed below. By contrast they include the new, compact Model 350; the big-wing, long-endurance Models 154 and 235; the Vietnam era 147T series; and the long-lived, reliable Model 124 (BQM) targets.

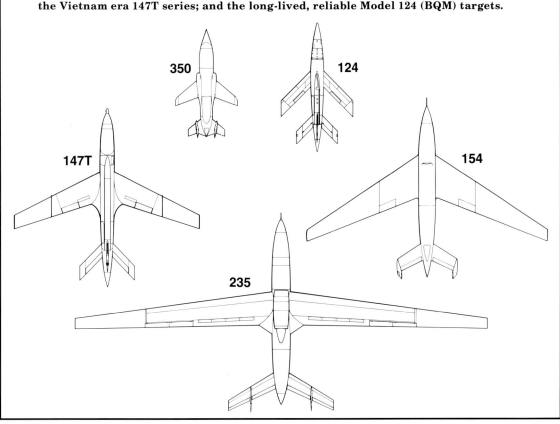

49	(Q-2, KDA, XM-21 series) 15,16	
124	(BQM-34A) **75-92**;	
	also see 16,20-22,65-74,96-99,104,168	
124E	(MQM-34D) . . . . .**84-87;** also see 66, 104	
124	(BQM-34AJ) **65-71**	
124I	(MQM-34D) . . . . **48-64**; *also see 135,159*	
124	(BQM-34S) **87,91-93**;	
	also see 139,140,148,150	
124RE	. . . .*(See Model 324)*	
147	(AQM series) . .**1-13**; *also see 100,104,108*	
147NA	(AQM-34H/J) . . .**102-105**; *also see 99,100*	
147SC	(AQM-34L/M) **104-108,142-150**;	
	also see 168	
147T	(AQM-34P/Q/R) 63,64,99,	
	100,108,110,115	
154	(AQM-91A) **27-47**;	
	also see 110-112,160,161,164	
166	(BQM-34E/F) **19-26**;	
	also see 63,90,92,100,120,168	
232	(BQM-34) **64,72-74**	
234	(BGM-34A/B) **98-101**;	
	also see 105,108	
235	(YQM-98A) **109-117**;	
	also see 155,160	

248	(BQM/SSM) **96-98**	
251	(MQM-34D Mod II) **86-87**	
255	(AQM-34V) **102-105**;	
	also see 96,108	
258	(ZBQM-111A) **119-123**	
259	(BGM-34C) . . . . .**105-108**; *also see 95,96*	
262	(STAR) Manta Ray **135,136**	
275	(Pre-production Compass Cope) 117	
305	(AQM-81A/N) . . . . .**124-129**; *also see 168*	
314	(Laser tests) **139,140**	
320	(YAQM-127A proposal) **130,131**	
324	(124RE) **157-174**;	
	also see 132,151,155,168,176-180	
326	(Helicopter RPV) **138**	
328	(Mini-RPV) **136,137**	
329	('Spirit' proposal) **134**	
336	(HATS) **140,141**	
350	(POC) . . . .**176-181**; *also see 14*	
350	(BQM-145A) **175-193**;	
	also see 14,132,151,155,156	
400	(Proposed Multi-National RPV) **152**	
410	(Affordable RPV) **151-156**;	
	also see 134,168,176,177	
410B	(POC) . . . .**176-181**; *also see 156*	

N

continued from page 200

National Association for Remotely Piloted
    Vehicles (NARPV) 17,18
Nationalist Chinese 35
NATO Missile Firing Installation (NAMFI)   50-52
Navajo Indian Reservation 42
NAVY
    Air Development Center 133-134
    Air Systems Command (NASC) 18,20,93,
            (*also NAVAIR*) 120,140,177,182
    Bureau of Aeronautics (Bu Aer) 22
    Bureau of Weapons (Bu Weps) 20-21
    Fighter Weapons School (Top Gun)
      Miramar Naval Air Station,
      San Diego 77-80,84
    Indian Head Arsenal 68
    Missile Test Center and Range . . . *see Pacific*
          Missile Test Center and Range
    Mobile Sea Range, Puerto Rico . . . 89,90,148
    Naval Ordnance Test Station (NOTS),
      China Lake, California 46,97
    Threat Simulation Department 23,80
Navstar Global Positioning System 134
New Mexico State Police 42
New York Times 49
Nietz, Tom 57
Nir, Maj. Shlomo 57,61
Nissho-Iwai 66,67
Nixon, President Richard M. 9,45-47,98,110
Nord Aviation 50,56
North American Aviation 37
North American Rockwell 93
North Atlantic Treaty Organization
    (NATO) 72,73,98
North Island Naval Air Station, San Diego . 22,89
North Warning System Radar (NWS) 142,145,147
Northrop and Northrop drones 17,37,61,176

O

100th Strategic Reconnaissance Wing
    (SRW) 4,45,108
Oemcke, Erich 6,21,116
Ogden, Utah 189, 190
O'Hara, Bruce 56
Okinawa 67,79
Oldfield, Frank 54,126,144,177
'Old Red' 82,128
Olmstead, Capt. John 182,188
Osama, Col. Elsayed Montasser, Egyptian
    Air Force 173
Osan, South Korea 9
Otchin, Arnie 160, 191
Over-the-Horizon Backscatter (OTH-B) . .**142-150**

P

P-3 launch plane 131
Pacific Missile Test Center and Range, Point
    Mugu, California 4,20-26,40-44,67,79-81,
        87-92,96,97,120,123,128,131,144,150

Packard, David 50
Palmdale, California 112
Panama, Canal Zone 84,85,144
Parachute Test Facility, El Centro, CA 31,40,136
Parafoil recovery system 183,184
Paris Air Show, France 159-161
Pathfinder 101
Patrick Air Force Base, Florida 115
Patriot anti-missile missile 67,87,120
Paul, Bernie 50,52,60
Pazmany, Ladislao 153
Peking, China 3,46
Peled, Gen. Benny 52,53
Pena Beach, Canal Zone 85
Perkins, Robert 54,63,64,71,73
Persian Gulf 87,90,139,148,150
Petrofsky, R.A. (Pete) 50,54,57,63,73
Philco-Ford 101
Philippine Sea, USS 145
Phoenix missile 80
Pico, Lou 104
Pilotless Target Aircraft (PTA) 64,**75-93**
Pitzen, Cmdr. John 79-80
Point Mugu, California . . . . . *see Pacific Missile*
          Test Center and Range
Popp, John 37
Port Said, Egypt 120
Porter, Ned 148
Portugali, Abraham and Phyllis 60
Powell, Col. Ellsworth A. 6
Powers, Francis Gary 1,49
Pratarelli, Corrado (Pat) 55,57
Precision Location Strike System (PLSS) . . . 111
Prisoners of War (POWs) 10
Proof-of-Concept (Model 350 prototype) . **177-181**
Pryor, Cliff 190
Puerto Rico . . . . *See Atlantic Fleet Weapons*
        Range and Roosevelt Roads
Pyramids (Egypt) 158,159,169,170

Q

Q-1, Q-2, Q-2C . . . . 15,16; *also see Firebee*
Quail missile 86
Quast, Capt. Phil M. 148

R

Radar Altimeter Control Equipment (RACE) . .58
Radar Altimeter Low Altitude Control
    System (RALACS) 26,97,120,140
Radar Target Scatter (RATSCAT) Facility,
    Holloman Air Force Base 43
Radioplane Company 15
Ranger (CV-61) USS 4
Rathgeber, Jack 52
Raynor, Col. Walter J. 7
Raytheon 72
Reasoner, Ron 21
'Red Book' 36
Red Sea .90
'Red Wagon' 1,36

Reda, Col. Mohamed Rashed, Egyptian
   Air Force 172,173
Regis, J.E. (Big Jim) 2
Reichardt, J.R. (Bob) 6,7,46,114
Remote Control Operator (RCO) 33,62,63
Republic Seabee 136
Requests for Proposal (RFPs) . . 116-118,126,130
RIMPAC (Pacific Rim) 89
Richards, A.C. (Tony) 176
Riley, Tom 179
Rockwell/Collins navigation 166
Rogers Lake, Edwards Air Force Base 114
Rogers, William, U.S. Secretary of State 48
Rommel, West German missile ship 25
Roosevelt Roads, Puerto Rico (Rosie Roads) 24,25,
    67,80,81,89-91,128,143,144,150
Royal Canadian Air Force (RCAF) 16
Rumpf, Richard 123
Russia 1,9,49,60,98
Russo-Japanese War (1904-5) 68
Rutan 'Varieze' plane 152
Rutan, Burt 162,166
Rutan/Yeager Flight 134
Rutherford, Amanda 158
Rutherford, Art 33,113
Rutherford, G. Williams (Bill) 36,37,51,58,
    93,158-160,164,169
Ryan Aeronautical Company 11,15
Ryan, General John D. 36,81
Ryan, Col. Lloyd M. 2,6,102
Ryan, T. Claude 11,36,176
'Ryan's Daughter' 6

S

6514th Test Squadron 25,45,99,100,
    103,106,142-150
71st Air Rescue Squadron 146,147
SA-2 and SA-3 Russian missiles . . . . *see Surface-*
    to-Air missiles
SH-60 Seahawk helicopter 183
Sahara Overseas Services 158,163
Saigon, South Vietnam 2,3,6
Sakamoto, Norman 111-117,121,131,
    161-167,177-178,181,182,185,186
Salt Lake City, Utah 18, 189, 190
Salto di Quirra Missile Range, Italy 72,74
San Clemente Island, California 98
San Juan Capistrano, California 139
San Juan, Puerto Rico 143
San Lorenzo, Sardinia, Italy 74
Santa Fe, New Mexico 33,34
Sardinia, Italy 72-74
Sargent, Ralph 68
Saudi Arabia 87
Savino, Mike 25,70,73,74,80
Scaled Composites, Inc. 134,162,166,178
Scarab Egyptian RPV (Model 324) **157-174**
Schwanhausser, Robert R. (Swany) . 1,2,6,21,33,
    36,39,46,50-52,59,60,98
Scud missile 87
Scurlock, Robert 163
Sealite 139
Sea of Galilee 52
Sea-skimmer 26
Sea Sparrow missile 120

Seattle, Washington 115
Second Fleet 79
Self-Propelled Air-to-Surface Missile
   (SPASM) 100
Semmes, Vice Adm. B.J., Jr. 79,90
Sevelson, Sam 20-22,120,131
Shapiro, Yacob, 'Shapik' 50,51
Sharif, Col., Egyptian Air Force 173
Sharp, Adm. U.S. Grant 11
Sheznai Hokido, Japan 67
Shillito, Barry J. 106,111,116
Shinto Shrine 69
Ship Tactical Airborne RPV (STAR) 135
Shohet, Lt. Aharon 61
Shulick, Capt. John 93
Sicily, Italy 74
Sidewinder missile 26,78,79,83,88-90
Signal Intelligence Gathering (SIGINT) . . . 110
Sikorsky CH-3E helicopter 38,43
Simone, John 138
Sinai, Israel-Egypt 54-60
Siren missile SS-N-9 120
Six-Day War 48-50,60,96,120
Sixth Fleet 74,90,96,148
'Skunkworks' 39
Sloan, William P. (Doc) 7,66,135
Sly, E.D. (Ed) 2,55,113-116,128
Smith, Charlie 21
Smith, Maj. Chuck 33
Smith, Lt. Col. Harold (Red) 17,102
Smith, Rear Adm. James H. 20
Smith, Cmdr. John C. 79-80
Smith, R.D. (Bob) 169,172
Smith, Richard E. 131
Soviet Union 50
Space and Naval Warfare Systems
   Command 140
Sparrow missile 24-26,78-80,88
Special Program Office (SPO) 17,39,110
Special Purpose Aircraft (SPA) 33,38,41-45
Sperry, Elmer 15
Sperry Univac 105
"Spirit of St. Louis" 2
'Spirit' proposal 132,134,162
STAR (Ship Tactical Airborne RPV) 135
St. Thomas, Virgin Islands 143
Standard Manufacturing 165
Standard missile 148
Star Wars 139,140
Stark, USS 139,148
Starkey, Hugh B. 105,108
Static Test Vehicle (STV) 40,42
Steakley, Brig. Gen. R.D. (Doug) 2,37
Stealth Technology . . . 1,36,38-39,112
Stewart, Lt. Gen. James T. 111,116,118
Stinger missile 86
Stockton, Bill 29
Stone, Elmer 66-67
Straits of Gibraltar 90,148
Strategic Air Command (SAC) . . . 2,4,36-39,45,
    46,102,104,108
Strategic Defense Initiative 140
Stubby Hobo missile 99,101
Styx missile (Russian) 96,120
Suez Canal 48-50,59-62,98
Super Q-2C 21
Supersonic Low-Altitude Target (SLAT) . . . 123,
    130,131

Surface-to-Air-Missile (SAM) 4,5,9,46-50,59,
　　　　　　　　　　　　　　　　63,98-100,103
Sved, B.J. (Billy) . . .2,85,89,100,106,136,143-150
Syria 50,61,63
System Qualification Test (SQT) 68

T

200th Drone Squadron 61
355th Tactical Wing 103
TACAN Guidance System 122
Tactical Air Command (TAC) 83,100-108,115
Tactical Landing Approach Radar
　(TALAR IV) 115
Taiwan 2,35
Takayama, Hideo 66-68
Talmor, Lt. Col. Uri 51-54,58
Talos missile 74,120,123
Tarter missile 67
Task Group 60.2 74
Tehachapi-Barstow, California 114
Tel Aviv, Israel 48-58,158,159
Teledyne, Inc. 11,158,164,176
Teledyne Brown Engineering 176
Teledyne Continental jet (CAE) and
　piston engines . . . . . see Continental listings
Teledyne Ryan Electronics 93
Teledyne Systems 40
Terrain Evade Procedure (TEP) 58
Terrier missile 24
Thailand 4
Thiokol rocket motors 120,121,166
Thomas, Dick 166
Thurmond, Senator Strom 108
Ticonderoga, USS 148
Tillman, Maj. Joe 108
Timmons, Gene 159-164
Timsa Lake, Suez Canal 60
Todd, R.L. (Bob) 55,57
Tokyo, Japan 66
Tomahawk cruise missile 18,121
'Tom Cat' 6
Tooele, Utah 189, 191
Top Gun School . see Navy Fighter Weapons School
Towbee System 67,84
Train, Vice Adm. Harry D., II 74
Tsiddon-Chatto, Yoash 50-52
Tsoor, Levi 52
Tucson, Arizona 47
Turkey 48
Tyndall Air Force Base, Florida . .25-26,48,52,77,
　　　　　　　　　　　　　　　80-83,88,128,144

U

UAV-MR　see Mid-Range Unmanned Aerial Vehicle
U-2 reconnaissance planes 1,35,48-50,104
Udall, Rep. Morris K. 31,32,44
United Nations 49,120
United Press International 30
United States Coast Guard 145,149
United Technology 126,131
　　　　　　　also see Chemical Systems Division
Unmanned Aerial Vehicle, Mid-Range
　(UAV-MR) 132,155,161,169
Utah Test and Training Range 143, 189-193

U-Tapao, Thailand 6,142
Utsunomiya, Japan 66
Uyehara, Steve 68,70

V

VA-34 Attack Squadron 11
VC-3 Composite Squadron 89
Vega Precision Systems 69,166
Vertical and Short Take-Off and Landing
　Aircraft (V/STOL) 46,110
Vertigo, Inc. 183
Vietnam (North and South) . . . 1-18,20,46,48,63,
　　　　　　　　　　　　78,79,84,92,98-104,
　　　　　　　　　　108,115,128,142,143,150
Vincennes, USS 148
VLF/Omega navigation 154
'Voyager' round-the-world flight 134,162

W

Wadi an-Natrum, Egypt 169-172
Wael, Col. A. El Maadawy, Egyptian Air Force 173
Wagner, Rear Adm. George F.A. 18,186
Wainwright, USS 24
Walker, Vice Adm. Thomas J. 24
Wall, Bill 54
Wallace Air Station, Philippines 78,88
Warner-Robins Air Force Base, Georgia 142
Weapon System Evaluation Program (WSEP) . .88
Weaver, Dale 2-6,33,42
Weaver, Keith B. (Buck) 113-116
Wendover Air Force Base, Utah 106,186
　　　　　　　also see Hill Air Force Base and
　　　　　　　　　　　　Dugway Proving Ground
West, Capt. Charles 5
White Sands Missile Range (WSMR),
　New Mexico 26,29-32,35,41,44,66,73,
　　　　　　　　77,80,84-87,128,139-141,165
White, Lt. Col. Wesley 149
William Tell Weapons Meet (Tyndall
　Air Force Base) 25,26,77,80-84
Williams, Cmdr. Bob 91
Wilson, Lt. Col Harold D. 25
Wilson, Teck A. 108,126
Wineteer, Capt. Larry D. 147,149
Witzel, Col. James 104,108
Woods, Col. Jim 113
Worth, Capt. Stan 33
Wright Brothers 15,126
Wright Field and Wright-Patterson
　Air Force Base, Ohio 17,37,104
Wright, Lt. Col. Richard 115
Wyman, Maj. John 33

X

X-13 Vertijet 110
XF2R-1 Dark Shark Fireball 110
XQM-103 (Flight Dynamics Laboratory
　Vehicle) 104
XV-5A/B Vertifan 110

Y

YAQM-127M (SLAT) 130,131
YQM-98A (Compass Cope-R) (Model 235) **109-117**
   Note: Y designates prototype
Yeager, Gen. Chuck 20
Yehuda, Master Sergeant 59
Yochim, Col. Fred 4
Yom Kippur War 47,61-64,135,159
Young, Jack 128
Yuma, Arizona 81

Z

ZBQ-90A 120
Zaki, Gen. Hagez N., Egyptian Air Force   161,173
Zangus, Lt. Col. Charles L. 24
Zelinski, Ray 169
Zeira, Maj. Gen. Eliahu 50
Zero Japanese Fighters 66